A CRACK IN CREATION

THE NEW POWER TO CONTROL EVOLUTION

Jennifer Doudna

Samuel Sternberg

THE BODLEY HEAD
LONDON

1 3 5 7 9 10 8 6 4 2

The Bodley Head, an imprint of Vintage,
20 Vauxhall Bridge Road,
London SW1V 2SA

The Bodley Head is part of the Penguin Random House group of companies
whose addresses can be found at global.penguinrandomhouse.com.

Penguin
Random House
UK

First published by The Bodley Head in 2017

www.vintage-books.co.uk

A CIP catalogue record for this book is
available from the British Library

Hardback ISBN 9781847923813
Trade paperback ISBN 9781847923820

Printed and bound by Clays Ltd, St Ives plc

Penguin Random House is committed to a sustainable future
for our business, our readers and our planet. This book is made
from Forest Stewardship Council® certified paper.

MIX
Paper from
responsible sources
FSC
www.fsc.org
FSC® C018179

To our parents, Dorothy and Martin Doudna (J.A.D.)
and Susanne Nimmrichter and Robert Sternberg (S.H.S.)

Science does not know its debt to imagination.

— Ralph Waldo Emerson

CONTENTS

PROLOGUE:
THE WAVE

IN MY DREAM, I am standing on a beach.

To either side of me, a long, salt-and-pepper strip of sand runs along the water, outlining a large bay. It is, I realize, the shore of the island of Hawaii where I grew up: the edge of Hilo Bay, where I once spent weekends with friends watching canoe races and searching for shells and the glass balls that sometimes washed ashore from Japanese fishing boats.

But today there are no friends, canoes, or fishing boats in sight. The beach is empty, the sand and water unnaturally still. Beyond the breakwater, light plays gently along the surface of the ocean, as if to soothe the fear I've carried since girlhood — the dread that haunts every Hiloan, no matter how young. My generation grew up without experiencing a tsunami, but we have all seen the photos. We know our town sits in the inundation zone.

As if on cue, I see it in the distance. A wave.

It is tiny at first but grows by the second, rising before me in a towering wall, its crest of whitecaps obscuring the sky. Behind it are other waves, all rolling toward the shore.

I am paralyzed with fear — but as the tsunami looms closer, my terror gives way to determination. I notice a small wooden shack behind me. It is my friend Pua's place, with a pile of surfboards scattered out front. I grab one and splash into the water, paddle out into the bay, round the breakwater, and head directly into the oncoming waves. Before the first one overtakes me, I'm able to duck through it, and when I emerge on the other side, I surf down the second. As I do, I soak in the view. The sight is amazing — there's Mauna Kea and, beyond it, Mauna Loa, rising protectively above the bay and reaching toward the sky.

I blink awake in my Berkeley, California, bedroom, thousands of miles away from my childhood home.

It is July 2015, and I am in the middle of the most exciting, overwhelming year of my life. I've begun having dreams like this regularly, and the recognition of their deeper meaning comes easily now. The beach is a mirage, but the waves, and the tangle of emotions they inspire — fear, hope, and awe — are only too real.

My name is Jennifer Doudna. I am a biochemist, and I have spent the majority of my career in a laboratory, conducting research on topics that most people outside of my field have never heard of. In the past half decade, however, I have become involved in a groundbreaking area of the life sciences, a subject whose progress cannot be contained by the four walls of any academic research center. My colleagues and I have been swept up by an irresistible force not unlike the tsunami in my dream — except this tidal wave is one that I helped trigger.

By the summer of 2015, the biotechnology that I'd helped establish only a few years before was growing at a pace that I could not have imagined. And its implications were seismic — not just for the life sciences, but for all life on earth.

This book is its story, and mine. It is also yours. Because it won't be long before the repercussions from this technology reach your doorstep too.

Humans have been reshaping the physical world for millennia, but the effects have never been as dramatic as they are today. Industrialization has caused climate change that threatens ecosystems around the globe, and this and other human activities have precipitated a surge in species extinction that is ravaging the diverse populations of creatures with which we share this earth. These transformations have prompted geologists to propose that we rename this era the Anthropocene — the human epoch.

The biological world is also undergoing profound, human-induced changes. For billions of years, life progressed according to Darwin's theory of evolution: organisms developed through a series of random genetic variations, some of which conferred advantages in survival, competition, and reproduction. Up to now, our species too has been shaped by this process; indeed, until recently we were largely at its mercy. When agriculture emerged ten thousand years ago, humans began biasing evolution through the selective breeding of plants and animals, but the starting material — the random DNA mutations constituting the available genetic variations — was still generated spontaneously and randomly. As a result, our species' efforts to transform nature were halting and met with limited success.

Today, things could not be more different. Scientists have succeeded in bringing this primordial process fully under human control. Using powerful biotechnology tools to tinker with DNA inside living cells, scientists can now manipulate and rationally modify the genetic code that defines every species on the planet, including our own. And with the newest and arguably most effective genetic engineering tool, CRISPR-Cas9 (CRISPR for short), the genome — an organism's entire DNA content, including all its genes — has become almost as editable as a simple piece of text.

As long as the genetic code for a particular trait is known, scientists can use CRISPR to insert, edit, or delete the associated gene in virtually any living plant's or animal's genome. This process is far simpler and more effective than any other gene-manipulation technology in existence. Practically overnight, we have found ourselves on the cusp of a new age in genetic engineering and biological mastery — a revolutionary era in which the possibilities are limited only by our collective imagination.

The animal kingdom has been the first and, so far, the biggest proving ground for this new gene-editing tool. For example, scientists have harnessed CRISPR to generate a genetically enhanced version of the beagle, creating dogs with Schwarzenegger-like supermuscular physiques by making single-letter DNA changes to a gene that controls muscle formation. In another case, by inactivating a gene in the pig genome that responds to growth hormone, researchers have created micropigs, swine no bigger than large cats, which can be sold as pets. Scientists have done something similar with Shannbei goats, editing the animals' genome with CRISPR so that they grow both more muscle (thus yielding more meat) and longer hair (meaning more cashmere fibers). Geneticists are even using CRISPR to transform Asian elephant DNA into something that looks more and more like woolly mammoth DNA, with the hope of someday resurrecting this extinct beast.

Meanwhile, in the plant world, CRISPR has been widely deployed to edit crop genomes, paving the way for agricultural advances that could dramatically improve people's diets and shore up the world's food security. Gene-editing experiments have produced disease-resistant rice, tomatoes that ripen more slowly, soybeans with healthier polyunsaturated fat content, and potatoes with lower levels of a potent neurotoxin. Food scientists are achieving these improvements not with transgenic techniques — the splicing of one species' DNA into a different species' genome — but by fine-tuned genetic upgrades involving changes to just a few letters of the organism's own DNA.

While applications in the planet's flora and fauna are exciting, it's the impact of gene editing on our own species that offers both the greatest promise and, arguably, the greatest peril for the future of humanity.

Paradoxically, some of the benefits to human health are likely to come from using CRISPR on animals or even insects. In recent experiments, CRISPR has been used to "humanize" the DNA of pigs, giving rise to hopes that these animals might someday serve as organ donors for humans. CRISPR is also tucked away inside the genomes of new mosquito strains, part of a plan to rapidly drive new traits into wild mosquito populations. Scientists hope to eventually eradicate mosquito-borne illnesses, such as malaria and Zika, or perhaps even wipe out the disease-carrying mosquitoes themselves.

To treat many diseases, though, CRISPR offers the potential to edit and repair mutated genes directly in human patients. So far we've gotten only a glimpse of its capabilities, but what we've seen in the past few years is exhilarating. In laboratory-grown human cells, this new gene-editing technology was used to correct the mutations responsible for cystic fibrosis, sickle cell disease, some forms of blindness, and severe combined immunodeficiency, among many other disorders. CRISPR enables scientists to accomplish such feats by finding and fixing single incorrect letters of DNA out of the 3.2 billion letters that make up the human genome, but it can be used to perform even more complex modifications. Researchers have corrected the DNA mistakes that cause Duchenne muscular dystrophy by snipping out only the damaged region of the mutated gene, leaving the rest intact. In the case of hemophilia A, researchers have harnessed CRISPR to precisely rearrange over half a million letters of DNA that are inverted in the genomes of affected patients. CRISPR might even be used to treat HIV/AIDS, either by cutting the virus's DNA out of a patient's infected cells or by editing the patient's DNA so that the cells avoid infection altogether.

The list of possible therapeutic uses for gene editing goes on and on. Because CRISPR allows precise and relatively straightforward DNA editing, it has transformed every genetic disease — at least, every disease for which we know the underlying mutation(s) — into a potentially treatable target. Physicians have already begun treating some cancers with souped-up immune cells whose genomes have been fortified with edited genes to help them hunt down cancerous cells. Although we still have a ways to go before CRISPR-based therapies will be widely available to human patients, their potential is clear. Gene editing holds the promise of life-changing treatments and, in some cases, lifesaving cures.

But there are other profound implications of CRISPR technology: it can be used not just to treat diseases in living humans but also to prevent diseases in future humans. The CRISPR technology is so simple and efficient that scientists could exploit it to modify the human germline — the stream of genetic information connecting one generation to the next. And, have no doubt, this technology will — someday, somewhere — be used to change the genome of our own species in ways that are heritable, forever altering the genetic composition of humankind.

Assuming gene editing in humans proves to be safe and effective, it might seem logical, even preferable, to correct disease-causing mutations at the earliest possible stage of life, *before* harmful genes begin wreaking havoc. Yet once it becomes feasible to transform an embryo's mutated genes into "normal" ones, there will certainly be temptations to upgrade normal genes to supposedly superior versions. Should we begin editing genes in unborn children to lower their lifetime risk of heart disease, Alzheimer's, diabetes, or cancer? What about endowing unborn children with beneficial traits, like greater strength and increased cognitive abilities, or changing physical traits, like eye and hair color? The quest for perfection seems almost intrinsic to human nature, but if we start down this slippery slope, we may not like where we end up.

The issue is this: For the roughly one hundred thousand years of modern humans' existence, the *Homo sapiens* genome has been shaped by the twin forces of random mutation and natural selection. Now, for the first time ever, we possess the ability to edit not only the DNA of every living human but also the DNA of future generations — in essence, to direct the evolution of our own species. This is unprecedented in the history of life on earth. It is beyond our comprehension. And it forces us to confront an impossible but essential question: What will we, a fractious species whose members can't agree on much, choose to do with this awesome power?

Controlling the evolution of the human species could not have been further from my mind in 2012, when my colleagues and I published the research paper that formed the basis of the CRISPR gene-editing technology. After all, our work had initially been motivated by curiosity about an entirely unrelated subject: the way that bacteria defend themselves against viral infection. Yet in the course of our research on a bacterial immune system called CRISPR-Cas, we uncovered the workings of an incredible molecular machine that could slice apart viral DNA with exquisite precision. The utility of this same machine to perform DNA manipulations in other kinds of cells, including human cells, was immediately clear to us. And as the technology was widely adopted and rapidly advanced, I could no longer avoid grappling with the numerous ramifications of our work.

By the time scientists had employed CRISPR in primate embryos to create the first gene-edited monkeys, I was asking myself how long it would be before some maverick scientists attempted to do the same in humans. As a biochemist, I had never worked with animal models, human tissues, or human patients; my comfort zone ended at the rims of the petri dishes and test tubes in my lab. Yet here I was, watching a technology I had helped create being used in ways that could radically transform both our species and the world in which we live. Would it inadvertently widen social or genetic inequalities or usher in a new eugenics movement? What repercussions would we need to prepare for?

I was tempted to leave those discussions to the people with actual bioethics training and get back to the exciting biochemical research that had drawn me to CRISPR in the first place. Yet at the same time, as a pioneer in the field, I felt a responsibility to help lead the conversation on how those technologies could, and should, be used. In particular, I wanted to ensure that the discussion involved not only researchers and bioethicists but also a great range of stakeholders, including social scientists, policymakers, faith leaders, regulators, and members of the public. Given that this scientific development affects all of humankind, it seemed imperative to get as many sectors of society as possible involved. What's more, I felt that the conversation should begin immediately, before further applications of the technology thwarted any attempts to rein it in.

And so, in 2015, while running my lab at Berkeley and traveling around the world to present my research at seminars and conferences, I began dedicating more and more of my time to subjects that were utterly foreign to me. I responded to dozens of reporters' inquiries about everything from designer babies to pig-human hybrids to genetically engineered superhumans. I spoke about CRISPR with the governor of California, with members of the White House Office of Science and Technology Policy, with the CIA, and in front of the U.S. Congress. I organized the first meeting to discuss the ethical questions that gene-editing technologies, and especially CRISPR, are raising in areas ranging from reproductive biology and human genetics to agriculture, the environment, and health care. And I helped build on the momentum from that first meeting by co-organizing a much larger international summit on human gene editing that brought together scientists and other participants from the United States, the UK, China, and around the world.

Again and again in these conversations, we have returned to the question of how this newfound power should be wielded. We have not yet arrived at an answer. But bit by bit, we are getting there.

Gene editing forces us to grapple with the tricky issue of where to draw the line when manipulating human genetics. Some people view any form of genetic manipulation as heinous, a perverse violation of the sacred laws of nature and the dignity of life. Others see the genome simply as software — something we can fix, clean, update, and upgrade — and argue that leaving human beings at the mercy of faulty genetics is not only irrational, but immoral. Considerations like these have led some to call for an outright ban on editing the genomes of unborn humans, and others to call for scientists to forge ahead without restraint.

My own views on the subject are still evolving, but I was struck by a comment made during the January 2015 meeting I organized to discuss human germline editing in embryos. Seventeen people, including the coauthor of this book (and my former PhD student), Sam Sternberg, were sitting around a conference table in California's Napa Valley having a heated debate about if and when germline editing could be allowed. Suddenly someone leaned into the group and said very quietly: "Someday we may consider it unethical *not* to use germline editing to alleviate human suffering." This remark turned our conversation on its head, and it still comes to mind whenever I meet with parents or would-be parents who are facing the devastating effects of genetic disorders.

While we deliberate, CRISPR research marches on. In mid-2015, Chinese scientists published the results of an experiment in which they had injected CRISPR into human embryos. The researchers had used discarded, nonviable embryos, but their study was nevertheless a major milestone: the first-ever attempt to precisely edit the DNA of the human germline.

There is justifiable alarm over developments like these. Yet we can't overlook the fantastic medical opportunities that gene editing gives us to assist people who suffer from debilitating genetic diseases. Imagine if someone who learned she carried the mutated copy of the *HTT* gene, which virtually guarantees early-onset dementia, had access to a CRISPR-based drug that could eliminate the DNA mutations before any symptoms appeared. Never before have curative treatments seemed so close, and it's essential that, as we debate germline editing, we take care not to turn public opinion against CRISPR or obstruct clinical uses of gene editing that are nonheritable.

I'm incredibly enthusiastic about the promise of gene editing. Progress in CRISPR research continues briskly in both academic labs and startup biotechnology companies, the latter supported by more than a billion dollars from investors and venture capital firms. Spurring the field, academic researchers and nonprofit groups are providing inexpensive, CRISPR-related tools to scientists around the world so that research can proceed unimpeded.

But scientific progress requires more than research, investment, and innovation; public involvement is also key. Up until now, the CRISPR revolution has taken place largely behind the closed doors of laboratories and biotech startups. With this book, as with other efforts, we hope to draw it into the light.

In part I of this book — "The Tool" — Sam and I share the thrilling backstory of CRISPR technology, including how it began with studies of a bacterial immune system and how it benefited from the decades-long journey to develop methods of rewriting DNA inside cells. In part II — "The Task" — we explore the myriad applications, both present and future, of CRISPR in animals, plants, and humans, and we discuss the exciting opportunities as well as the significant challenges that lie ahead. You'll notice that we use my voice throughout. Both of us wrote the book, and both of us share most of the views expressed herein. But we took this narrative approach for the sake of clarity and to capture the breadth and details of my unique experiences over the years.

This book is not intended to be a rigorous history of CRISPR or an exhaustive chronology of gene editing's early development. Rather, we have tried to highlight some of the most pertinent advances and provide glimpses of how our own work dovetailed with others' research. We have included references where appropriate, and we encourage interested readers to consult additional publications to supplement the topics we discuss. Finally, we humbly acknowledge the countless scientists who have played crucial and invaluable roles in the study of CRISPR and gene editing, and we apologize to the many colleagues whose work we didn't have space to mention.

We hope this book will demystify this exciting area of science and inspire you to get involved. A global discussion about gene editing has already begun; it's a historic debate about nothing less than the future of our world. The wave is coming. Let's paddle out and ride it together.

Part I
THE TOOL

1

THE QUEST FOR A CURE

RECENTLY, I HEARD an incredible story, one that sums up the power and promise of gene editing.

In 2013, scientists at the National Institutes of Health were grappling with a medical mystery. These researchers were studying a rare hereditary disease known as WHIM syndrome and had come across a patient whose condition they simply could not explain. Early in life she had been diagnosed with the disorder, yet when the NIH scientists met her, it seemed to have miraculously vanished from her system.

Affecting just a few dozen people worldwide, WHIM is a painful, potentially deadly immunodeficiency disease that makes life difficult for those unfortunate enough to suffer from it. It is caused by a tiny mutation — a single incorrect letter among some six billion total letters of one's DNA, amounting to a change of just a dozen or so atoms. This minute transformation leaves WHIM victims profoundly susceptible to infection by human papillomavirus (HPV), which causes uncontrollable warts that cover the patient's skin and can eventually progress to cancer.

It's a testament to the rareness of the disease that the patient in whom WHIM syndrome had first been diagnosed back in the 1960s was the same person whom the NIH researchers met all those years later. In the scientific literature, she's known simply as WHIM-09, but I'll call her Kim. Kim had been afflicted with WHIM since birth, and over the course of her life, she had been hospitalized multiple times with serious infections stemming from the disease.

In 2013, Kim — then fifty-eight — presented herself and her two daughters, both in their early twenties, to the staff at NIH. The younger women had classic signs of the disease, but the scientists were surprised to discover that Kim herself seemed fine. In fact, she claimed to have been symptom-free for over twenty years. Shockingly, and without any medical intervention, Kim had been cured.

Scientists conducted experiments to understand how Kim had spontaneously recovered from her life-threatening illness, and they found some telling clues. The mutated gene responsible for Kim's condition was still present in cells taken from her cheek and skin, but in her blood cells, the mutation was inexplicably absent. Analyzing the DNA taken from Kim's blood cells in more detail, the scientists found something even more extraordinary: One copy of chromosome 2 was missing a whopping thirty-five million letters of DNA, a section that included the entirety of the mutated gene, called *CXCR4*. (Gene names are written in italics; the proteins they code for are in regular typeface. For example, the *HTT* gene codes for a protein called huntingtin; Huntington's disease is caused by a mutation in the *HTT* gene.) The roughly two hundred million letters of DNA that remained of chromosome 2 were scrambled; it was as if a tornado had swept through the chromosome and left its components in complete disarray.

These initial findings raised a host of new questions. How had the DNA in Kim's blood cells become so irregular when the DNA was normal (aside from the *CXCR4* mutation) everywhere else in her body? Moreover, given that the chromosome harboring the *CXCR4* gene was so badly damaged, with 164 genes now missing, how were the blood cells not only still alive but functioning normally? The human genome — the complete set of all the genetic information in our cells — contains thousands of genes that are required for vital functions, such as DNA replication and cell division, and it seemed almost inconceivable that so many genes could simply vanish without any harmful consequences.

After running a battery of tests, the NIH scientists slowly pieced together an explanation for Kim's serendipitous cure. They concluded that a single cell in her body must have experienced an uncommon and usually catastrophic event called chromothripsis — a recently discovered phenomenon in which a chromosome suddenly shatters and is then repaired, leading to a massive rearrangement of the genes within it. The effects in the body are generally either trivial (if the damaged cell dies immediately) or dire (if the rearranged DNA inadvertently activates cancer-causing genes).

In Kim's body, though, chromothripsis turned out to have quite another effect. Not only did the mutated cell grow normally, but — because it was now rid of the diseased copy of *CXCR4* — the cell was free of the gene causing WHIM syndrome.

But Kim's blind luck had not ended there. The NIH scientists determined that the fortunate cell must have been a hematopoietic stem cell, a type of stem cell from which every kind of blood cell in the body originates and that has an almost unlimited potential to proliferate and self-renew. That cell had passed along its rearranged chromosome to all its daughter cells, eventually repopulating Kim's entire immune system with healthy new white blood cells that were free of the *CXCR4* mutation. This chain of events — so unlikely that I had a hard time even conceiving of it as I listened to the scientist's presentation — had effectively wiped out the disease that had haunted Kim since birth.

As the researchers who studied Kim's condition wrote in their summary of her case, Kim was the beneficiary of an "unprecedented experiment of nature" in which a single stem cell underwent a spontaneous change that rid the cell, and all its progeny cells, of a diseased gene. It was, simply put, a blessed accident — one that, had it unfolded differently, could have killed Kim but that instead had arguably saved her life.

To understand just how fortuitous this outcome was, imagine that the human genome is a large piece of software. In Kim's case, the software contained one letter of faulty code among the approximately six billion that made up the program. To troubleshoot the problem, you wouldn't just go around randomly deleting large chunks of code and scrambling other parts. Not only would that almost certainly fail to correct the original mistake, but you'd most likely introduce other, bigger problems in your blind attempts to correct the error. Only if you were incredibly lucky — managing to beat million-to-one, or even billion-to-one, odds — would you happen to both delete a chunk containing the misspelled code *and* do it in a way that didn't destroy the critical function of the software. In a nutshell, that's what happened in Kim's genome — except the blind programmer in this instance was nature itself.

But as incredible as Kim's case is, what's even more amazing is that she's not alone. While hers is the only reported case of a patient being cured by spontaneous chromosome shattering and repairing, the scientific literature is peppered with other examples of patients who were partially or completely cured of a genetic disease through accidental, spontaneous "editing" of the genome. For example, in the 1990s, two New York patients were diagnosed with a genetic disorder called severe combined immunodeficiency (SCID), also known as the "bubble boy" disease because of the sterile environments in which some children have been contained to reduce their exposure to pathogens. Without extreme isolation or aggressive forms of therapy, patients with SCID usually die before they're two years old. Yet the two SCID patients in New York were the exceptions to this terrible rule; they remained surprisingly healthy into adolescence and adulthood. The reason in both cases, scientists determined, was that the patients' cells had spontaneously corrected the disease-causing mutation in a gene called *ADA*, and they'd done it without disturbing the remainder of the gene or the chromosome.

Similar cases of natural gene editing have cured other genetic diseases, such as Wiskott-Aldrich syndrome (a disorder from which a whopping 10 to 20 percent of patients are saved by spontaneous genetic correction) and a liver condition called tyrosinemia. In certain skin diseases, the presence of gene-edited cells is visible to the naked eye. The evocatively named condition ichthyosis with confetti, for instance, leaves victims with patches of red and flaky skin. The cells in these areas carry a genetic mutation, but the cells in the surrounding, healthy patches of skin have managed to fix the mutation.

Overall, however, the odds of being spontaneously cured of a genetic disease are minuscule. Most patients will never experience the natural miracle of having their genomes altered in exactly the right way, in the right kinds of cells, and in the right tissues. Natural gene editing remains an anomaly — some interesting medical cases involving a handful of patients who won the genetic lottery, nothing more.

But what if gene editing weren't only a spontaneous event? What if doctors had a way to fix the harmful mutations that cause WHIM syndrome, SCID, tyrosinemia, or, for that matter, any other genetic disease?

For many scientists, myself included, cases such as Kim's were exciting not only because they revealed the curative power of natural gene editing, but also because they shone a light on a potential avenue of medical intervention: a way of reversing the effects of genetic disease by rationally and deliberately correcting misspellings in the genome. These good-luck stories demonstrated that intentional gene editing would be possible if scientists had the genetic know-how — and the biotechnological tools — to pull it off.

For decades, long before I entered the field, women and men in the life sciences had labored to gain this know-how and develop these tools. In fact, scientists were dreaming of therapeutic gene editing long before they were aware that nature had provided the clues to create it. To make this sort of technology possible, however, researchers needed to understand the genome itself: what it was made of, how it was built, and — most important — how it could be modified and manipulated. Only with that basic knowledge could they and their scientific descendants take the first, halting steps toward helping people who, unlike Kim, were powerless to cure themselves.

The genome — a term coined in 1920 by the German botanist Hans Winkler and probably intended as a portmanteau of *gene* and *chromosome* — refers to the entire set of genetic instructions found inside a cell. Mostly identical from cell to cell within any given organism save for the occasional mutation, the genome tells all living things how to grow, how to sustain themselves, and how to transmit genes to offspring. One organism's genome directs it to grow fins and gills that let it move and breathe underwater; another organism's genome instructs it to produce leaves and chlorophyll that let it harvest energy from sunlight. Our intrinsic physical traits — eyesight, height, skin color, predisposition to disease, and so on — are the result of information encoded in our genomes.

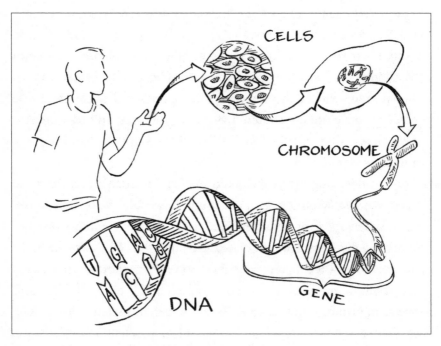

DNA: *the language of life*

The genome is made up of a molecule called deoxyribonucleic acid, or DNA, which is constructed of just four different building blocks. Known as nucleotides, these are the familiar letters of DNA: A, G, C, and T, shorthand for the chemical groups (also known as bases) of adenine, guanine, cytosine, and thymine that distinguish the four compounds. The letters of these molecules are connected in long single strands. Two of these strands come together to form the famous double-helix structure of DNA.

The double helix is a bit like a ladder twisted into a long, spiraling coil. Two strands of DNA wrap around each other along a central axis, with the continuous sugar-phosphate backbone of each one occupying the outside of the helix; together, these form the two side rails of the ladder. This arrangement positions the four different bases on the inside of the helix, projecting inward and meeting up in the middle; these are the rungs of the ladder. An elegant feature of the structure is the set of chemical interactions that hold the two strands together at each rung, sort of like molecular glue: The letter A from one strand always pairs with T on the other strand, and G always pairs with C. These are known as base pairs.

The double helix beautifully reveals the molecular basis of heredity; this is how a relatively simple chemical like DNA can transmit genetic information to two daughter cells upon cell division, and it is how that information can be further propagated to every cell of an entire plant or animal. By virtue of the molecule's two-strandedness and the rules governing how those strands assemble (A with T, G with C), each strand acts as a perfect template for its matching pair. Shortly before cell reproduction, the two strands are separated by an enzyme that "unzips" the double helix right down the middle. After that, other enzymes build a new partner strand for each single strand simply by using the same base-pairing rules, resulting in two exact copies of the original double helix.

My own introduction to the DNA double helix coincided with my

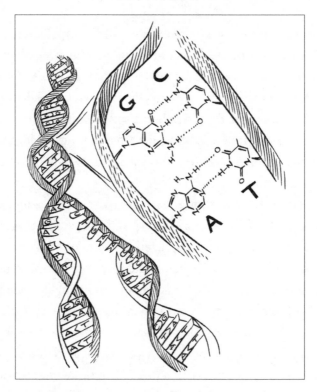

The structure of the DNA double helix

discovery that scientists could learn about molecules that were too tiny to see with even the most powerful light microscopes. I came home from school one day when I was about twelve or so to find a tattered copy of James Watson's *The Double Helix* lying on my bed. (My dad would occasionally pick up books for me at used bookstores to see if they sparked any interest.) Thinking this book was a detective novel — which it was! — I set it aside for some weeks before diving into its pages one rainy Saturday afternoon. Reading Watson's account of the incredible academic collaboration with Francis Crick that had enabled them — using crucial data collected by Rosalind Franklin — to discover this simple and beautiful molecular structure, I felt the first tugs of interest that would eventually guide me onto a similar path. (Many years later, I would jump-start my own scientific career by determining some of the first three-dimensional structures of far more complex RNA molecules.)

In the years that followed Watson and Crick's discovery, scientists sought to understand how this molecule's structure and rather simple chemical ingredients could encode information and explain the multitudinous phenomena of biological life. DNA, as it turns out, is much like a secret language; each specific sequence of letters provides instructions to produce a particular protein inside the cell. The proteins then go on to carry out most of the critical functions in the body, like breaking down food, recognizing and destroying pathogens, and sensing light.

To transform the instructions contained in DNA into proteins, cells use a crucial — and closely related — intermediary molecule called ribonucleic acid, or RNA, which is produced from the DNA template via a process called transcription. RNA has three of the same letters as DNA, but in RNA, the letter T (for thymine) is replaced with the letter U (for uracil). In addition, the sugar that makes up the backbone of RNA contains one more oxygen atom than the sugar in DNA (hence the name *deoxy*ribonucleic acid). RNA acts as a messenger, ferrying information from the nucleus, where the DNA is stored, to the outer regions of the cell, where proteins get produced. In a process called translation, cells use long strings of RNA that are produced by discrete segments of DNA — stretches of code called genes — to construct individual protein molecules. Every three letters of RNA, when read together, equal one amino acid, and amino acids are the building blocks of proteins. Genes and their corresponding protein products differ from one another by the sequence of nucleotides (in genes) and amino acids (in proteins). This overall flow of genetic information — from DNA to RNA to protein — is known as the central dogma of molecular biology, and it is the language used to communicate and express life.

The size of a genome and the number of genes it contains varies widely across different kingdoms of life. Most viruses, for instance, have just a

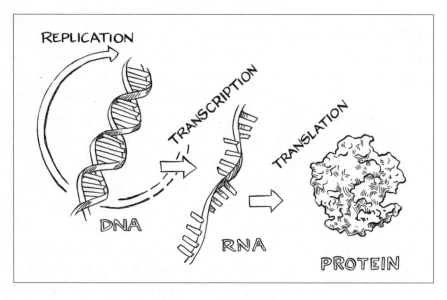

The central dogma of molecular biology

few thousand letters of DNA (or RNA, since some viral genomes contain no DNA) and a small handful of genes. Bacterial genomes, by contrast, are millions of letters long and contain around 4,000 genes. Fly genomes contain around 14,000 genes spread out across hundreds of millions of DNA base pairs. The human genome comprises about 3.2 billion letters of DNA, with around 21,000 protein-coding genes. Interestingly, a genome's size is not an accurate predictor of an organism's complexity; the human genome is roughly the same length as a mouse or frog genome, about ten times smaller than the salamander genome, and more than one hundred times smaller than some plant genomes.

Different species package their genomes in vastly different ways. Whereas most bacterial genomes exist inside the cell as a single continuous piece of DNA, the human genome is composed of twenty-three distinct pieces, called chromosomes, that range in length from 50 to 250 million letters. Like the cells of almost all mammals, human cells normally contain two copies of each chromosome, one from the father, one

itated by new technology that allowed researchers to clone large chunks of human DNA in yeast, as well as by major upgrades in laboratory automation and developments in complex computational algorithms that helped parse the sequencing data. In 2001, after herculean efforts and at a cost of more than three billion dollars, the first draft of the genome was published.

Since the completion of the Human Genome Project, the process of DNA and whole-genome sequencing has become staggeringly quick, cheap, and effective. Scientists have precisely identified well over four thousand different kinds of DNA mutations that can cause genetic disease. DNA sequencing can tell individuals if they're at elevated risk of developing certain cancers, and it can help tailor specific treatments to best match the genetic backgrounds of different patients. Furthermore, now that commercial DNA sequence analysis has gone mainstream and costs just a few hundred dollars per test, millions of individuals have opted to have their own genomes analyzed by simply dropping a saliva sample in the mail. The resulting explosion of data has helped researchers pinpoint significant associations between thousands of gene variants and a number of physical and behavioral traits.

Yet, while genome sequencing represents a huge development in the study of genetic disease, it is ultimately a diagnostic tool, not a form of treatment. It has allowed us to see how genetic diseases are written in the language of DNA, but it leaves us powerless to change that language. After all, it's one thing to learn how to read; it's quite another to learn how to write. For that, scientists need an entirely different set of tools.

Researchers have been dreaming of DNA-based disease cures for as long as they've known about genetic disease. Even as some scientists were beginning to locate the root causes of genetic disorders, others were aggressively pursuing new techniques for treating those afflictions — not just by giving patients drugs to temporarily alleviate the adverse effects of a mutated gene, but by repairing the gene itself to permanently reverse the course of the disease. To take just one, sadly common example: Sickle cell disease is treated with frequent blood transfusions, the use of the drug hydroxyurea, and bone marrow transplants. Wouldn't it be better to target the causative DNA mutation itself?

The best solution for treating genetic disease, these early researchers knew, would be to fix the defective gene, doing purposely what nature had done accidentally when it cured Kim and other lucky patients like her. For these scientists, however, the idea of treating genetic disease by rewriting the mutated genetic code seemed impossible. Repairing a defective gene would be like finding a needle in a haystack and then removing that needle without disturbing a single strand of hay in the process. However, they suspected that they could effect a similar change by adding entire replacement genes to damaged cells. The question was, how could they deliver that precious cargo to an ailing genome?

Inspired in part by the uncanny ability of viruses to splice new genetic information into the DNA of bacterial cells, the pioneers of this early gene therapy realized they could use viruses to deliver therapeutic genes to humans. The first reported attempts came in the late 1960s from Stanfield Rogers, an American physician who had been studying a wart-causing virus in rabbits, Shope papillomavirus. Rogers was particularly interested in one aspect of the Shope virus: It caused rabbits to overproduce arginase, an enzyme their bodies used to neutralize arginine, a harmful amino acid. The sick rabbits had much more arginase in their systems, and much less arginine, than healthy rabbits. What's more, Rogers found that researchers who had worked with the virus *also* had lower-than-normal levels of arginine in their blood. Apparently these scientists had contracted the infections from the rabbits, and these infections had led to lasting changes in the researchers' bodies as well.

Rogers suspected that the Shope virus was ferrying a gene for heightened arginase production into cells. As he marveled at the virus's ability to transfer its genetic information so effectively, he began to wonder if an engineered version could deliver other, useful genes. Many years later,

Rogers would recall: "It was clear that we had uncovered a therapeutic agent in search of a disease!"

Rogers didn't have to wait long for a disease on which to test his theory. Just a few years later, a genetic disorder called hyperargininemia was discovered in two German girls. Like the rabbits infected with Shope papillomavirus, these patients had abnormal levels of arginine — but rather than having unusually low levels of the amino acid, their levels were extremely high. The patients' gene for arginase production — the gene that Rogers suspected was transmitted by the Shope virus — was either missing or mutated.

The symptoms of hyperargininemia are awful; they include progressively increasing spasms, epilepsy, and severe mental retardation. But there was a chance that early intervention, especially in the younger of the two German patients, might stave off the worst effects of the disease. Rogers and his German collaborators administered Shope to the girls therapeutically, injecting large doses of the purified rabbit virus directly into their bloodstreams.

Unfortunately, Rogers's experimental gene therapy was a disappointment, not only for him but also, and especially, for his patients and their families. The injections had little effect on either girl, and Rogers himself was widely censured for performing a procedure that many scientists considered reckless and premature. Later research would demonstrate that, in contrast to Rogers's theory, the Shope virus didn't even contain an arginase gene and so would not have been useful for treating hyperargininemia in the first place.

Although Rogers would never attempt gene therapy again, his approach of using viruses as gene-delivery vehicles — vectors, as scientists call them — revolutionized the field of biology. The experiment failed, but its basic premise proved solid, and viral vectors are still one of the most effective ways we know to insert genes into a cell's genome and, thus, alter the genetic code of living organisms.

A few specific traits make viruses effective as vectors. For starters, vi-

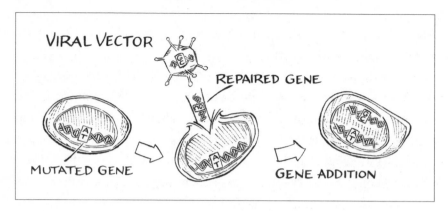

Gene therapy using viral vectors

ruses have evolved to be incredibly efficient at infiltrating cells of any type. For as long as life has existed, organisms from all kingdoms — bacteria, plants, animals, and so on — have had to contend with parasitic viruses, whose sole goal is to hijack cells, insert their own DNA into them, and trick the cells into creating more copies of the virus. Over the eons, viruses have learned how to exploit practically every weak spot in a cell's defense system, and they have perfected strategies of dumping their genetic payload into the interior of the cell. As a tool, viral vectors are astoundingly reliable; researchers working with viral vectors can get genes into target cells with nearly 100 percent efficiency. For the scientists who pioneered their therapeutic use, viral vectors were the ultimate Trojan horse.

Viruses know not only how to get their DNA inside a cell but also how to make the new genetic code stick. In the 1920s and 1930s, the early days of bacteria-focused genetics research, scientists were puzzled by the ability of bacterial viruses to emerge seemingly out of nowhere and cause infections. Subsequent research demonstrated that these viruses could actually splice their genome into the bacterial chromosome and lurk there, undetected, until the conditions were just right to initiate an aggressive infection. Retroviruses, a large class of viruses that includes the human immunodeficiency virus (HIV), do the same thing in humans, splicing their genetic material into the genome of infected cells. This pernicious property makes retroviruses especially challenging to eradicate, so much so that they have left an outsize mark on our species. A full 8 percent of the human genome — over 250 million letters of DNA — is a remnant of ancient retroviruses that infected ancestors of our species millennia ago.

Following the first attempts at gene therapy in the 1960s, the field took off, thanks to a revolution involving recombinant DNA — a catch-all term for genetic code produced in a lab, not by nature. Using new biotechnology tools and new biochemical methods, scientists in the 1970s and 1980s developed ways to cut and paste segments of DNA into genomes and isolate specific gene sequences. This enabled them to insert therapeutic genes into viruses and remove dangerous genes so that the viruses would no longer harm infected cells. Scientists had essentially turned these viruses into benign missiles, engineered to deliver their genetic payload to the desired target, but little more.

By the late 1980s, after retooled retroviruses had been successfully used to insert lab-produced genes into mice, the race was on to test gene therapy in the clinic. I was at Harvard at the time, conducting research for my PhD in biochemistry, and I recall discussing with my lab mates the news that French Anderson and his colleagues at the National Institutes of Health had been first to the finish line. They developed a promising vector doped with a healthy copy of the adenosine deaminase (*ADA*) gene, which is mutated in patients suffering from ADA-SCID (severe combined immunodeficiency). Their goal was to use gene therapy to permanently integrate the nonmutated *ADA* gene into the blood cells of ADA-SCID patients, allowing them to produce the missing protein — a step that Anderson and his colleagues hoped would cure the disease. Unfortunately, the results of this pioneering clinical trial were murky; the retooled virus didn't harm either of the two recipients, but its effectiveness was hard to determine. For instance, both patients had increased levels of viable immune cells following the procedures, but that improvement could have been attributable to other therapies they were receiving concurrently. What's more, because only a very low number of the patients' cells seemed to have actually received the healthy *ADA* gene, it appeared that the virus might not have been as efficient at gene splicing as scientists had hoped.

Since that early, inconclusive test nearly three decades ago, however, gene therapy has seen some phenomenal advances. Improvements in the design of viral vectors and in the methods used to deliver them have led to extremely encouraging results for *ADA* gene therapy in dozens of SCID patients, and a commercially available version of it, Strimvelis, is likely to be approved soon. What's more, in the two thousand or so gene-therapy trials that have been completed or initiated as of 2016, the list of targetable conditions has expanded dramatically and now includes other monogenic heritable diseases, like cystic fibrosis, Duchenne muscular dystrophy, hemophilia, some forms of blindness, and a growing number of cardiovascular and neurological diseases. Meanwhile, the budding field of cancer immunotherapy — in which tumor-fighting cells are loaded with genes that target molecules specific to tumors — has been hailed as one of the most promising breakthroughs for cancer treatment and proof that gene therapy still has much to contribute to the field of medicine.

But despite the hype, gene therapy hasn't been the panacea that scientists and physicians had hoped it would be; in fact, at times it seems to have done more harm than good. The field received a shock in 1999 when a patient died after suffering a massive immune response to a high dose of viral vector. By this time I was a faculty member at Yale University and deeply engaged in projects to determine how viral RNA molecules hijacked the protein-making machinery in cells. Although my field of study was far removed from that of gene therapy, news of this disastrous outcome made me both sad and determined to work toward a deeper understanding of viruses and cells.

Then, in the early 2000s, five patients in a gene therapy trial for

X-linked SCID developed leukemia — a cancer of the bone marrow. The cancers resulted from the retrovirus's errant activation of an oncogene — a cancer-causing gene — which caused the cells to proliferate uncontrollably. This incident underscored the inherent risks of giving patients large quantities of a foreign agent and randomly jamming a few thousand letters of DNA into their genomes. I remember thinking to myself that this line of clinical research, so exciting in principle, just seemed too inherently risky.

Gene therapy, by its very nature, is also ineffective for a wide range of genetic conditions that aren't caused by missing or deficient genes. Such conditions can't be fixed by simply delivering new genes into cells. Take Huntington's disease, in which the altered gene produces an abnormal protein that completely overrides the effect of the second, healthy copy of the gene. Since the mutated gene dominates the nonmutated gene, simple gene therapy — the addition of another normal copy of the gene using a retooled virus — would have no effect on Huntington's or other dominant conditions.

For these and many other hard-to-treat genetic diseases, what doctors really needed was a way to repair problematic genes, not just supplant them. If they could fix the defective code that caused the issue, they could target recessive and dominant diseases alike without ever having to worry about the consequences of splicing a gene into the wrong place.

This possibility had intrigued me since the beginning of my career. In the early 1990s, after leaving Harvard with my PhD, I discussed this very idea on many evenings in the lab at the University of Colorado, Boulder, where I was working as a postdoctoral researcher. In those days, my friend and lab mate Bruce Sullenger and I debated everything from the 1992 presidential election — I liked Paul Tsongas, Bruce liked Bill Clinton — to various strategies for gene therapy. One idea we kicked around was the possibility that RNA molecules, those intermediaries between DNA and proteins in cells, could be edited to fix mutations they carried over from DNA. This was in fact the subject of Bruce's research project.

Occasionally we also discussed another possibility: editing the source code of such defective RNAs — that is, the actual DNA of the genome. This would be game changing, we agreed. The question was, would it ever be anything but a pie-in-the-sky idea?

Throughout the 1980s, as some researchers refined virus-based gene transfer therapies, others pursued simpler methods of transforming mammalian cells, using DNA prepared in the lab. These basic methods were intended mostly for research, but as the decade progressed, scientists began to explore their therapeutic potential in human cells as well.

These approaches possessed a few key advantages over more complicated gene transfer techniques. For starters, they were much quicker; rather than going to all the trouble of packaging genes inside rebuilt viruses, scientists could inject their lab-made DNA directly into cells or allow cells to soak it up spontaneously by bathing them in a specially prepared mixture of DNA and calcium phosphate. Second, although these simpler methods didn't involve virus-assisted gene splicing into the cells' genome, the cells were able to merge the foreign DNA with their own — albeit inefficiently.

Mice were often the first test subjects for these techniques, and scientists were amazed at how effective the new methods were in the tiny creatures. By injecting new DNA into fertilized mouse eggs and then implanting those eggs into female mice, researchers found they could permanently splice the foreign DNA into the next generation and cause observable changes in the developing animals. These advances meant that any gene that scientists could isolate and clone in the lab could be tested and investigated; by adding the gene to cells, scientists could observe its effects and better understand the gene's function. Although my own research at the time focused on figuring out the shapes and functions of RNA molecules, I could appreciate that the implications were enormous.

The question was, how exactly was DNA finding its way into the genome? Mario Capecchi, a professor at the University of Utah, pursued this problem in the early 1980s after making the puzzling observation that when many copies of a gene were spliced into the genome, the pattern of integration was the opposite of the randomness that one might have expected. Instead of the gene copies being distributed haphazardly throughout the different chromosomes of the genome, Capecchi found that the genes were always clustered together in one or a few regions, with many copies overlapping one another, as if they'd been deliberately assembled. In fact, he determined, that's exactly what had happened.

Capecchi had observed the effects of a process called homologous recombination — a well-known phenomenon by that time, but not one he'd expected to see in this experiment. Homologous recombination occurs most famously during the formation of egg and sperm cells, when the two sets of chromosomes we inherit from our parents are pared down to just one, to be combined with a second set during sexual reproduction. In this process of elimination, cells select a blend of the paternal and maternal chromosomes; each pair of chromosomes engage in their own version of sex, exchanging large chunks of DNA in a way that increases genetic diversity. Despite the mind-boggling complexity of mixing, matching, and reassembling millions of letters of DNA, cells can do this flawlessly using homologous recombination. The same process occurs in all kingdoms of life; bacteria, for instance, exchange genetic information through homologous recombination, and biologists have been taking advantage of homologous recombination to conduct genetics experiments in yeast for years.

But Capecchi's discovery that mammalian cells cultured in the lab also engaged in homologous recombination was momentous. As he mentioned at the end of his 1982 article, "It will be interesting to determine whether we can exploit [the enzymes involved] to 'target' a gene by homologous recombination to a specific chromosomal location." Homologous recombination, in other words, might allow scientists to precisely paste genes into matching sites in the genome — a dramatic improvement over the randomness of gene splicing with viruses. Better yet, it might enable scientists to overwrite defective genes simply by plugging in healthy replacements directly at the site of the mutation.

Just three years after Capecchi's study, that possibility became a reality with a notable paper published by Oliver Smithies and his colleagues. Working with human cells derived from bladder tumors, they set out to replace the cells' homegrown copies of the beta-globin gene with an artificial, recombinant version that had been constructed in the lab. Unbelievably, it worked. Without the scientists' needing to play any fancy tricks — they literally just mixed the DNA with calcium phosphate and squirted it on the cells — a few of the cells internalized the foreign DNA, paired up the DNA sequences built in the lab with their matching DNA sequences in the genome, and then performed some molecular gymnastics to swap out the old and replace it with the new.

Cells, it seemed, could do most of the hard work of modifying their genomes all by themselves. This meant that scientists could deliver genes more gently, without using viruses to ram new DNA into the genome. By tricking a cell into thinking that the recombinant DNA was simply an extra chromosome that needed to be paired with a matching gene already in its genome, scientists could ensure that the new DNA was combined with the existing, native genetic code through homologous recombination.

Scientists dubbed this new approach to gene manipulation gene targeting. Today, we know it by another name: gene editing.

The potential of such technology for genetics research was tantalizing. But Smithies knew that homologous recombination could also be used as a therapy. If scientists could perform similar gene targeting in the blood stem cells of a patient suffering from sickle cell disease, the mutated beta-globin gene could be replaced with the normal, healthy sequence. His discovery, still just an experimental approach, might someday be used to cure disease.

Other laboratories scrambled to refine this gene-targeting technique. One of them was Capecchi's. In 1986, while I was in my second year

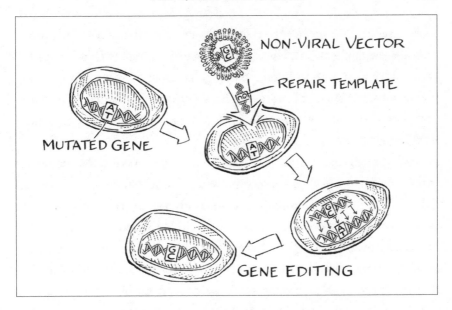

NON-VIRAL VECTOR

REPAIR TEMPLATE

MUTATED GENE

GENE EDITING

Gene editing via homologous recombination

of graduate school, he demonstrated that homologous recombination was precise enough to fix even single mutations in the genome and to correct enzyme deficiencies in cells. Two years later, he presented a general-purpose strategy for targeting any gene in the genome for which the underlying sequence was known. He also proposed that homologous recombination could be used not just to correct or repair genes, but also to inactivate them for research purposes; by switching genes off and observing the results, scientists could discern the genes' functions.

By the time I completed my PhD at the end of the 1980s, gene targeting had been widely used to edit DNA in cultured mouse and human cells and even in live mice. Seminal work in Martin Evans's lab demonstrated that by targeting genes in mouse embryonic stem cells and then injecting those modified stem cells back into mouse embryos, scientists could create live mice with designer changes. The breakthroughs by Capecchi, Smithies, and Evans were eventually recognized with the 2007 Nobel Prize in Physiology or Medicine.

Despite its earth-shaking implications, however, gene editing in its early days had far more appeal for basic research than it did for human therapeutic applications. For mammalian geneticists looking for ways to study the functions of different genes, gene targeting was a game-changing technology. But medical researchers were wary of using it in humans, for despite its potency, when it came to treatment, homologous recombination fell pathetically short.

Perhaps its biggest drawback was the problem of nonhomologous recombination, or illegitimate recombination, in which the new DNA was haphazardly integrated into the genome rather than accurately delivered to its matching sequence. In fact, illegitimate recombination seemed to prevail over homologous recombination by a factor of about a hundred to one. Clearly, therapeutic use wasn't very promising if gene editing could correct a mutated gene in only 1 percent of transformed cells while randomly splicing DNA into the genome in the other 99 percent. Scientists developed elegant workarounds to circumvent the problem in cell cultures, and they didn't lose hope for the future application to medicine. As Capecchi stated in the early 1990s, "Eventually, homologous recombination for human gene therapy is the only way to go." But for the time being, it seemed, gene editing just wasn't good enough to use on humans.

In the early 1980s, while many scientists were busy thinking about targeting genes in human cells, Jack Szostak had been puzzling over the process of yeast cell division. A professor at Harvard Medical School (who would later mentor me during my PhD research project), Szostak was occupied with the fundamental problem of how gene targeting and homologous recombination were even possible. Specifically, he wanted to understand how two strands of DNA from one chromosome could merge with two matching strands of DNA from a second chromosome, exchange information during some kind of fused intermediate stage, and then separate again to re-form the individual chromosomes after cells divide.

In 1983, while I was still an undergraduate at Pomona College, on the other side of the country, Szostak thought he'd found an answer. Based on results from yeast genetics experiments, he and graduate student Terry Orr-Weaver, along with Professors Rodney Rothstein and Frank Stahl, published a provocative model in which the precipitating factor — the red alarm that initiated the process of homologous recombination — was one of the two chromosomes being sliced apart, causing a DNA double-strand break. In this model, a double-strand break and the freed ends of DNA at the site of the break would be particularly prone to fusing, with the flanking sequences far likelier to engage in the exchange of genetic information with a matching chromosome (or, in the case of gene editing, with the matching DNA supplied by the researcher).

By the time I arrived in his lab in 1986, Szostak was already switching his research focus to the role of RNA molecules in the early evolution of life. But in the lab, my peers and I discussed the double-strand-break model and its elegance, as well as the frank skepticism with which it had been received in the scientific community. But over time, it became clear that this model was consistent with extensive experimental data. The double-strand-break repair mechanism made sense not only for the homologous recombination that occurred during the formation of egg and sperm cells, but also for the recombination that occurred whenever DNA was damaged. All cells are exposed to DNA-damaging agents, such as x-ray radiation and carcinogens, and cells are remarkably efficient at repairing those breaks without losing genetic information. According to Szostak's model, that repair process depended on the ability of chromosomes to match up via homologous recombination, which might be why having two copies of a chromosome was a beneficial evolutionary strategy. Any damage to one chromosome could be repaired simply by copying the matching sequence on the second chromosome.

If the double-strand-break model was correct, and if the conclusions from yeast research held true for mammals, then there was an obvious opportunity to improve the efficiency of gene editing: slice apart the ge-

nome precisely where editing was being attempted. If you wanted to replace a defective gene in the genome with a corrected copy that had been
constructed in the lab, you would first have to figure out how to cut the
defective gene apart, inducing a local double-strand break in the DNA,
and then provide the corrected gene copy. Faced with a break, the cell
would attempt to repair the damage by searching for a matching chromosome to copy — at which point it would find the synthetic gene. Essentially, you'd trick the cell into thinking it had suffered a natural source
of DNA damage and supply it with a new piece of DNA, disguised as a
second chromosome, that it could use to fix the broken site.

Researchers in Maria Jasin's laboratory at Memorial Sloan Kettering
Cancer Center in New York City were the first to play this game of de-
ception on mammalian cells in 1994 — a development I read about with
keen interest from nearby New Haven, where I had just arrived after fin-
ishing up my postdoctoral research in Boulder. It was thrilling to learn
of this pioneering work, built on the double-strand-break model of my
graduate school adviser, being done by another female scientist who
shared my fascination with the molecules of life.

Jasin's gene-editing experiment was original and inventive. Her strategy was to introduce an enzyme into mouse cells that sliced the genome
apart, making a double-strand break; at the same time, she'd add a piece
of synthetic DNA to the cells — a repair template — that matched the
DNA sequence that had been cut. Later she would check to see if the
mouse cells had repaired the broken DNA by plugging in the repair
template. By running the same experiment with and without the added
enzyme, she could test her hypothesis: that an artificially generated double-strand break enhanced the efficiency of homologous recombination.

The challenge was coming up with a viable enzyme that would cut the
genome in one specific place out of billions of possible options. To solve
that problem, Jasin cleverly stole a piece of molecular machinery from
yeast: the I-*Sce*I endonuclease.

Nucleases are enzymes that cut apart nucleic acids; some cut RNA, others cut DNA. Endonucleases cut RNA or DNA somewhere *within* the strands, as opposed to exonucleases, which cut exclusively from the ends. Some endonucleases are highly toxic to cells because they cut just about any piece of DNA they find, regardless of its sequence. Other endonucleases are highly specific and cut only certain sequences, and many more fall somewhere in the middle.

The I-*Sce*I endonuclease that Jasin selected was one of the most specific endonucleases known at the time, requiring a perfect match of eighteen consecutive DNA letters for it to cut a given segment. Selecting a highly discriminating endonuclease was critical; if Jasin had chosen an enzyme that was too promiscuous, it would cut the genome all over the place, not only making the results more difficult to interpret, but potentially harming the host cell. With specificity for eighteen letters in a row, though, I-*Sce*I would cut just one sequence of DNA out of the more than fifty billion possible combinations. (Ironically, the mouse genome didn't even have a matching eighteen-letter sequence, so before attempting the gene-editing experiment, Jasin had to splice a copy of the sequence into the genome so the enzyme would have somewhere to cut.)

The results of Jasin's experiment were astounding. She succeeded in inducing a whopping 10 percent of the cells to precisely repair a mutated gene by homologous recombination, a success rate that seems low now but that was hundreds of times higher than what scientists had managed to achieve previously. It was the most promising evidence yet that this process might allow scientists to rewrite the code of the genome without the risk of illegitimate recombination or random splicing from retroviral vectors. Introduce a double-strand break in the right place, and cells would practically do the work for you.

There was just one problem: For this approach to be useful, scientists had to be able to cut the genome in specific locations. In Jasin's proof-of-concept experiment, the sequence recognized by I-*Sce*I had been artificially glued into the genome before the nuclease was introduced, but the sequences of many disease-associated genes were set in stone, so to speak; they couldn't be modified to match some persnickety endonuclease enzyme. Once broken, the genome was highly effective at repairing itself and incorporating new genetic information — the trick was to figure out how to break it in the right place.

From the mid-1990s onward, while I was delving into the molecular structures of RNA molecules and their unique biochemical behaviors, researchers rushed to design new systems that, like I-*Sce*I, could accurately target specific DNA sequences. If scientists could solve that problem, they would be able to unlock the full potential of gene editing.

These next-generation gene-editing systems had three critical requirements: They had to recognize a specific, desired DNA sequence; they had to be able to cut that DNA sequence; and they had to be easily reprogrammable to target and cut different DNA sequences. The first two criteria were necessary for generating a double-strand break, and the third was necessary for the tool to be broadly useful. I-*Sce*I excelled at the first two but failed miserably at the third. To build a programmable DNA-cutting system, bioengineers figured they would need to either re-tool I-*Sce*I to target and cut new kinds of sequences or find a completely new nuclease enzyme that had already evolved to cut different DNA sequences.

Scientists' efforts to redesign I-*Sce*I fell short (not surprising, given the sheer molecular complexity of protein enzymes), and it quickly became apparent that mining nature for other nuclease enzymes would be a much more promising approach. In fact, by the time of Jasin's work with I-*Sce*I, scientists had already isolated dozens of nucleases from a wide range of organisms and determined the exact DNA sequences that they targeted. But there was a fundamental problem: the vast majority of these enzymes recognized sequences that were only six or eight letters long — far too short to be useful. Those sequences occurred tens of thousands or even hundreds of thousands of times in the human genome, meaning that even if the nuclease could stimulate homologous recombination in one gene, it would shred up nearly the entire genome in the process. The cell would be destroyed before it ever had a chance to initiate DNA repair.

Researchers couldn't rely on any of the previously discovered nucleases, and it wasn't feasible to go searching for new enzymes like I-*Sce*I every time a new gene edit was desired. If therapeutic gene editing were to be a viable technique for correcting disease-causing mutations, physicians couldn't wait around for scientists to discover a nuclease that happened to target the precise region of the exact gene where a patient had a harmful mutation. Scientists needed to be able to pluck the correct nuclease off the shelf or at least have a way of generating it on demand.

Although I was unaware of it at the time, a paradigm-shifting study that presented a solution to this problem took place in 1996. Srinivasan Chandrasegaran, a professor at Johns Hopkins University, realized that instead of building nucleases from scratch, finding new ones in nature, or remaking I-*Sce*I, he could take a hybrid approach by selecting pieces of proteins that existed naturally and combining them. Such chimeric nucleases would fulfill the first two requirements of a gene-editing nuclease: they would be able to recognize and cut a specific sequence of DNA.

Chandrasegaran set about piecing together a chimeric nuclease using sections from two naturally occurring proteins that were adept at, respectively, targeting and cutting sequences of DNA. To do the cutting, Chandrasegaran selected a module from a bacterial nuclease called FokI that could introduce breaks in DNA but had no particular sequence preference. To do the targeting, he harnessed a family of ubiquitous, naturally occurring proteins called zinc finger proteins, so named because they recognized DNA using fingerlike extensions held together by zinc ions and arranged side by side, just like the fingers of a hand. Because these zinc finger proteins were built of multiple repeated segments arranged in tandem, with each segment recognizing a specific three-letter

DNA sequence, it seemed likely that scientists could redesign the proteins to recognize various DNA sequences by combining segments in different ways.

Incredibly, Chandrasegaran's chimeric nuclease seemed to work. After fusing the cutting module from FokI with the DNA-recognition module from a zinc finger protein, his team demonstrated that the designer nuclease recognized and cut exactly the DNA they'd expected it to, even though they had mashed together two protein components from completely different sources.

Soon, Chandrasegaran teamed up with University of Utah professor Dana Carroll to put these new zinc finger nucleases, or ZFNs, to more practical use. Together, they demonstrated that ZFNs also worked in frog eggs (a popular model system for biologists), and that ZFN-induced DNA cutting stimulated homologous recombination. Next, working in fruit flies, Carroll's lab programmed a new ZFN to target a gene involved in body pigmentation called *yellow* and showed that this strategy could produce a precise genetic alteration in a whole organism. This was a profoundly significant development for gene editing. Not only were ZFNs practical enough to use in animals but, more important, they could be redesigned to target new genes.

The broader scientific community quickly jumped on board, and researchers began designing ZFNs for their own individual purposes, targeting new genes and working in new model organisms. In 2003, Matthew Porteus and David Baltimore were the first to show that a gene in human cells could be precisely edited by a custom-built ZFN. Soon thereafter, Fyodor Urnov and colleagues corrected a mutation causing X-linked SCID in human cells. The possibility of using a gene-editing strategy to target genetic diseases had never been more real.

Meanwhile, ZFNs were also adopted by labs that were interested in gene editing for completely different purposes, such as producing precisely engineered crops or animal models. In the late 2000s, the technology was successfully applied to thale cress, tobacco plants, and corn,

demonstrating that DNA double-strand breaks promoted highly efficient homologous recombination in many cell types, not just mammalian. Concurrently, papers reporting that ZFNs had been used to modify
genes in zebrafish, worms, rats, and mice began trickling out. This work
was intriguing and caught my attention in publications and at conferences due to its exciting potential.

But in spite of their promise, ZFNs were never widely adopted outside of a handful of labs. The researchers who used them had extensive
experience in protein engineering, collaborations with the few labs that
already had such experience, or deep pockets with which to pay the hefty
price for the designer nucleases. In theory, designing ZFNs was easy—
simply combine the different zinc finger segments in such a way that
they would recognize the DNA sequence you were interested in editing.
But in practice, it was very difficult. A high proportion of newly designed
ZFNs simply didn't recognize the DNA sequences they were supposed
to; others were too promiscuous and went after anything remotely related to their targets, killing the cells they were tasked to edit. In still
other cases, the zinc finger module might recognize the DNA just fine,
but the nuclease module wouldn't cut.

For some of the same reasons that retooling I-SceI had proved challenging, it seemed, ZFNs just weren't quite programmable enough to be
useful as a multipurpose gene-editing tool. To be sure, the results with
ZFNs had proved conclusively that designer nucleases were the way to
go when gene editing was the objective, but the field was still waiting for
a new kind of technology, one that would be more reliable and easier to
use.

That technology—at least, the first version of it—was discovered in
2009 and came from studies of novel types of proteins found in *Xan-
thomonas,* a pathogenic plant-infecting bacterium. Called transcription
activator–like effectors, or TALEs, these proteins are remarkably similar
to zinc finger proteins in their construction: they're built out of multiple
repeating segments in which each segment recognizes a given area of

DNA. But there is a difference: whereas each finger of the zinc finger proteins recognizes a three-letter sequence of DNA, each segment in TALEs recognizes just a single letter of DNA. This difference allowed scientists to easily deduce a code for what segment would recognize a given letter of DNA, and then they simply arranged those segments, one after the other, to recognize a longer sequence of DNA within a gene. This had *seemed* straightforward for ZFNs, but with TALEs, it actually was.

Scientists quickly pivoted to explore this latest lead. Soon after this code was discovered, three laboratories fused TALEs to the same DNA-cutting module used in ZFNs and created TALE nucleases, or TALENs. TALENs were remarkably effective at initiating gene editing inside cells, and after researchers made some improvements in their design and construction, TALENs, it seemed, would be far easier to build and implement than ZFNs.

"But pity the poor TALENs," as Dana Carroll wrote in an article chronicling gene editing's origins. For no sooner had TALENs been discovered and adapted for gene editing than they were supplanted by the next, and possibly ultimate, arrival in the gene-editing field. The technology was called CRISPR, and it's here that my story dovetails with that of gene editing — and with this long march of scientific history, which was about to enter an exhilarating new phase.

2

A NEW DEFENSE

IN 2014, I celebrated the twenty-year anniversary of my research laboratory — a milestone that happened to coincide with my fiftieth birth-
day — by organizing a lab retreat to my childhood home: Hawaii. The
thirty-odd people who attended (a combination of undergraduates, PhD
students, postdoctoral scientists, lab personnel, significant others, and
even my son, Andrew) stayed in three rental properties near the town of
Kona, on the western shore of the Big Island, just fifteen minutes from
the beach and a few hours' drive from the house in Hilo where I grew
up. During the day, we picnicked, hiked through Hawai'i Volcanoes National Park, trekked to the nearby beaches and markets, and snorkeled
amid the pristine reefs surrounding the island. We spent one spectacular
evening enjoying breathtaking views of the red aura created by natural lava flows in the Halema'uma'u Crater, and several nights socializing
over pizza and beer back at our rented houses, letting down our hair with
spontaneous dance parties and karaoke sessions.

Of course, as with any scientific get-together, we also made time for
presentations. Over the course of four days, we held four mini-symposia
in which each member of the lab gave a fifteen-minute talk on a topic of
his or her choice, the subjects ranging from the lab's history to the finer
points of RNA structure.

On the fourth day, Ross Wilson, a postdoctoral scientist who had arranged most of the retreat's logistics, got up to give the last talk. At least,

I thought it would be a talk. Instead, Ross surprised us all with a short movie of clips of me that he'd spliced together from a box of old VHS tapes that — unbeknownst to me — had been passed down over the years as a sort of lab tradition.

The guests cheered and ribbed me in equal measure as one clip after another appeared on the screen: video from an acceptance speech I had made at a National Science Foundation awards ceremony back in 1999, a shot of me holding a Geiger counter in *Vogue* magazine in 2000, and an excerpt from a documentary that Frederick Wiseman had filmed in my lab, which, by this time, had moved from Yale to the University of California, Berkeley.

Buried in this footage were snippets of two television news stories I had appeared in, both of them covering the first major discovery to come out of my Yale lab back in 1996. I remembered the existence of these TV pieces, if not their particulars. The sudden rush of attention on my lab had been both exciting and a bit unnerving, especially for a young researcher who spent most of her time sequestered away at the laboratory bench.

Of all the clips in Ross's video, these were the ones that provoked the loudest hoots from the group. It all seemed so retro — their boss in her thirties, the old-fashioned tone of the newscasters, and the spectacle of the clunky, now-obsolete computers that had been state-of-the-art back in the day.

As I joined in the laughter, my mind glided back through the years to those early days of my work at Yale, and I remembered the hopes and fears that I had faced as I embarked on a risky new area of research, a project that many scientists had warned me would never pan out. Watching my younger self being interviewed for these news segments brought back the feelings of both intense exhilaration and profound loss that had colored those years. My recorded comments also provided a surprising forecast of what was to come much later as my research progressed in new directions.

At the time of the interviews, my lab had just determined the three-dimensional structure — the precise location of every single atom — of a molecule of ribonucleic acid, or RNA, which formed part of a larger molecule called a self-splicing ribozyme. In the 1980s, Tom Cech, my postdoctoral adviser at the University of Colorado, Boulder, had been awarded the Nobel Prize for his discovery of self-splicing ribozymes. His discovery was a breakthrough because the existence of self-splicing ribozymes suggested that life on earth arose from molecules of RNA that could both encode genetic information and replicate that information in primitive cells. When I started my lab at Yale in 1994, I had aimed to build on Tom's breakthrough by studying the structure of the ribozyme to better understand how it worked. I wanted to figure out how RNA — a molecule closely related to DNA — could function as both a repository of genetic instructions and a chemically reactive molecule capable of changing its shape and biological behavior. This effort had culminated in the fantastically exciting discovery that RNA can fold up into three-dimensional structures that are very different from the elegantly simple DNA double helix.

But my joy in determining the ribozyme structure, work I did with graduate student Jamie Cate, was accompanied by personal tragedy. That fall, my father called my office at Yale with terrible news: he had been diagnosed with advanced melanoma. During the last three months of his life, I flew to Hawaii from New Haven three times and spent intense days and nights holding his hand, reading him favorite passages from Henry David Thoreau's *Walden*, listening to Mozart, discussing how different pain medications worked, and considering what happens to us after death. Always interested in what I was researching, Dad continually asked about my latest results from the lab. At one point, I showed him a picture of the ribozyme molecule, rendered in green ink. "Looks like green fettuccine!" he said. Three weeks later, he was gone.

Reeling from my father's death but longing for distraction, I threw myself back into my work, taking comfort in the thought that other people's lives might one day be saved, or at least improved, by our research. The ribozyme project, like much scientific research, was driven by two desires: to shed light on unexplored natural phenomena, and to put this knowledge to practical use. Back when I had decided to determine the molecular structure of the ribozyme, many biologists thought that this type of molecule might offer an alternative method for treating diseases. The ribozyme-based method, as it was envisioned at the time, would differ from both gene therapy (which aimed to fix genetic mishaps by adding healthy genes) and gene editing (which aimed to repair the defective genes themselves) in that it would allow clinicians to cure patients by fixing defective RNA molecules — those messengers that our cells use to convert DNA into protein.

In my excitement about our ribozyme breakthrough, I speculated to the TV interviewer that these molecules might one day become tools for editing DNA. After all, there was already evidence that some ribozymes were capable of triggering chemical changes in DNA. Watching the nearly twenty-year-old video footage, I saw my younger self make a beeline for that very application: "One possibility," I had said, "is that we might be able to cure or treat people that have genetic defects . . . We hope that [this discovery] will provide some clues as to how we might be able to modify the ribozyme so that it can act like a molecular repair kit and repair defective genes."

As it turned out, this particular development never came to pass, or at least it hasn't yet. While a number of ribozyme-based therapies eventually made their way into clinical trials, none proved effective for treating genetic diseases. But the interview jolted me back to the present with its unexpected connection to my current research.

What caught my attention as I sat in that rental house in Hawaii was how the words I had chosen back then reflected a surprising twist in the path of my work. When I described our ribozyme research in terms of its potential for repairing genes, I had no idea that, nearly two decades later, gene editing would come to define my career.

Around fifteen years after those news segments aired, I participated in a line of investigation whose therapeutic promise was far, far greater than anything I had imagined as a new faculty member in 1996. It happened while I was studying another biological system, an immune system in bacteria, in which RNA played a starring role. But unlike the ribozyme project, which focused on a topic that had already received a huge amount of attention due to its discoverer's Nobel Prize, this journey began in obscurity. It had started as a lark and progressed through a series of unlikely meetings and serendipitous collaborations. Sitting there in Hawaii with family and coworkers, watching my younger self on TV, I marveled at how the underlying idea of repairing defective genes seemed to be threaded through my career.

I will never forget the first time I heard the term *CRISPR*.

It was 2006, and I was sitting in my office on the seventh floor of Stanley Hall at the University of California, Berkeley, when the phone rang. On the line was Jillian Banfield, a fellow Berkeley professor from the Departments of Earth and Planetary Science and Environmental Science, Policy, and Management.

I knew Jill only by reputation, and she knew equally little about me; she explained that she had found my lab's website only after a quick Google search. A geomicrobiologist who focused primarily on the interactions between microorganisms and their environments, Jill had been looking for faculty members at Berkeley who were researching RNA interference, a molecular system that plant and animal cells use to suppress the expression of particular genes and that organisms also use during their immune responses. It was a subject with which my lab had extensive experience.

Jill told me that her lab was studying something I heard as *crisper* — she didn't define or even spell the term, only mentioned that it had popped up in some data sets her lab was analyzing — and that she wanted to expand her investigation using tools from genetics and biochemistry, two skill sets my lab could offer. In particular, she thought there might be some parallels between "crisper" and RNA interference. Would I like to meet and discuss it?

I was intrigued by Jill's intensity, although I had my doubts about her request; I had no idea what she was studying, after all. But her excitement was palpable even over the phone, so I agreed to meet her for coffee the following week.

After the call, I did a quick search in the scientific literature and found only a handful of articles on the subject Jill had been so excited about. By comparison, RNA interference, the study of which was scarcely eight years old, had already racked up over four thousand references. (The attention would reach a climax when its discoverers, Andrew Fire and Craig Mello, won a Nobel Prize later that year.) The relative dearth of publications on Jill's topic made it challenging to evaluate — but it also piqued my curiosity.

I skimmed several of the background articles, reading just enough to learn that this thing — CRISPR — referred to a region of bacterial DNA and that the acronym stood for "clustered regularly interspaced short palindromic repeats." I didn't read any further, stymied by the jargon and figuring that Jill would fill me in when we met.

My own Google search showed just how accomplished a scientist Jill was. Brilliant and creatively engaged in multiple diverse areas of science, she'd published articles with titles such as "Mineralogical Biosignatures and the Search for Life on Mars" and "Geophysical Imaging of Stimulated Microbial Biomineralization." Her research involved collecting and studying samples from far-flung sources, like biospheres deep in the earth beneath Japan, hypersaline lakes in Australia, and acidic mine drainages in Northern California. These exotic projects stood in marked contrast to my own; aside from requiring frequent trips to the x-ray-generating particle accelerator at the Lawrence Berkeley National Laboratory, my lab's work took place mostly inside test tubes.

Partly because I was so impressed with her research and partly for my own scientific reasons, I was growing keener to meet Jill. I had relocated to Berkeley from Yale University four years earlier with my now-husband, Jamie Cate, and our newborn son, Andrew. Although my ongoing research had moved in some new directions, I was hoping to expand the lab and pick up some additional projects while also cultivating partnerships with new colleagues. This could be just the lead I was looking for.

Jill and I met the following week at the Free Speech Movement Café, near the entrance to one of the campus's undergraduate libraries. It was a blustery spring day, and when I arrived, Jill was already seated at a stone table in the outdoor courtyard, a notepad and a stack of papers beside her. After we had chatted for a bit, she picked up her notebook and got down to business.

She quickly sketched a diagram of CRISPR. First, she drew a large oval to represent a bacterial cell. Then she drew a circle inside the oval, the bacterial chromosome, and added a series of alternating diamonds and squares along one side of the circle to represent a region of DNA. This region, apparently, was CRISPR.

Jill shaded in the diamonds and mentioned that they were all identical stretches of the same thirty-ish letters of DNA. Then she numbered the squares sequentially beginning at 1 while explaining that each one constituted a unique sequence of DNA.

Finally, the words that the acronym stood for — *clustered regularly interspaced short palindromic repeats* — began to make sense to me. The diamonds were the short repeats, the squares were interspacing sequences that regularly interrupted the repeats, and these diamond-square arrays were clustered in just one region of the chromosome, not randomly distributed throughout. (When I inspected the repeating DNA sequences more closely, back in my office, the *P* in the acronym also became clear: the sequences were nearly the same when read in either direction, just like a palindrome such as "senile felines.")

The basic idea that cells could contain repeating DNA sequences was not itself a surprise; more than 50 percent of the human genome — well

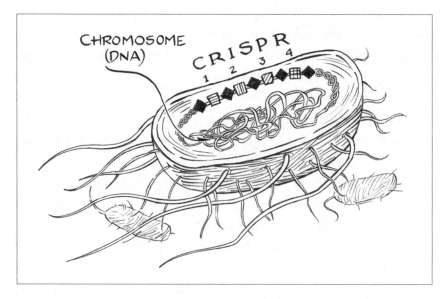

CRISPR inside a bacterial cell

over one billion letters of DNA — comprises different types of repetitive arrays, some of which are copied millions of times. Although comparatively smaller bacterial genomes contain far less, I was aware that they too had repeating sequences, some of which shared terms with CRISPR, like repetitive extragenic palindromic (REP) sequences and bacterial interspersed mosaic elements (BIME). But I had never heard of DNA repeating itself with this kind of precision and uniformity, where every repetition was truly identical and always separated from its neighbor by a similarly sized, random spacer sequence.

Curious to learn more about these strange regions of bacterial DNA, I asked Jill about their biological function and was disappointed to hear that she didn't know. But her lab had uncovered an important clue: DNA sequences from natural bacterial populations showed that essentially every cell had a different CRISPR array due to the unique sequences interspaced between the repeats. This was completely unprecedented because every other part of the DNA was nearly identical in each of these cells. Jill recognized that CRISPRs were probably the fastest-evolving regions in the genome, consistent with a function that had to change or adapt quickly in response to something the cells encountered in their environment.

Years earlier, pioneering work by Francisco Mojica, a professor in Spain, had uncovered the same kinds of repeats in numerous completely unrelated species, including in archaea, single-celled microorganisms that, like bacteria, lack nuclei. (Bacteria, Archaea — referred to collectively as prokaryotes — and Eukarya constitute the three domains that encompass all life on earth.) CRISPRs, Jill said, had been found in almost half of all the bacterial genomes that had been sequenced to date and in nearly every archaeal genome. In fact, they appeared to be the most widely shared family of repeating DNA sequences in all prokaryotes.

These bits of information sent a little shiver of intrigue down my spine; if CRISPR was present in so many different species, there was a good chance that nature was using it to do something important.

I listened attentively as Jill pulled three scientific publications, all from 2005, out of her stack of papers and excitedly summarized the results. Working independently, three research laboratories, including the one headed by Mojica, had found that many of the CRISPR spacers — those snippets of DNA sandwiched between the repeating sequences — were perfect matches with the DNA of known bacterial viruses. Even more intriguing, there seemed to be an inverse correlation between the number of DNA sequences in a bacterium's CRISPR that matched viral DNA, and the number of viruses that could infect the CRISPR-containing bacterium; the more matches, the lower the threat of infection. Jill's own pioneering research, in which DNA genomes from entire microbial communities were reconstructed by sequencing smaller overlapping segments of DNA and assembling them together, also showed that many of the interspaced sequences in the CRISPR arrays matched viral DNA sequences in the environment.

Together, the findings provided a major hint about the role CRISPR plays in bacteria and archaea. These teams of researchers had uncovered evidence suggesting that CRISPR was likely to be part of an archaeal and bacterial immune system, an adaptation that allowed microbes to fight off viruses.

Last, as if laying out the final bread crumb, Jill showed me the latest article about CRISPR. Published by a National Institutes of Health team led by Kira Makarova and Eugene Koonin, it was called "A Putative RNA-Interference-Based Immune System in Prokaryotes"—a title that immediately hooked me. Although this article, like the previous three, lacked conclusive experimental data, its authors had done an impressive job of synthesizing the available information about CRISPR. Combining the results of a number of earlier studies with an expert analysis of the prevalence of CRISPRs in different species, they had pieced together an intriguing new hypothesis that suggested RNA was a key participant in the immune system of single-celled microorganisms like bacteria — and that this system might be functionally similar to one of my research interests, RNA interference.

Jill couldn't have chosen better bait to lure me into her investigation. Not only had I spent my entire career up to that point studying RNA molecules, but I had become increasingly focused on the process of RNA interference in human cells. Now, Makarova and Koonin were suggesting that CRISPR was bacteria's equivalent to RNA interference. If that was true, then my lab was perfectly situated to tackle this new, mysterious biological function. The prospects were all the more tantalizing because, while other scientists had floated hypotheses about CRISPR, no one had yet conducted the experiments to prove or disprove those theories. The timing couldn't have been better for biochemists like me to jump into the fray and start figuring out how CRISPR worked.

As Jill and I parted ways, I thanked her and promised to stay in touch. I would need to mull over all this information and think about the costs and benefits of adding CRISPR research to my lab's workload. If I did take it on, I'd have to find a scientist to run the program on a day-to-day basis, since I was too busy managing the lab to tackle a new project by myself.

I would also need to do some studying to brush up on the world of bacteria and the viruses that infect them. I had published numerous journal articles on the hepatitis C virus, I was studying the influenza virus with a new postdoc in my lab, and I knew that the RNA interference pathway was intimately connected to the antiviral defenses of plants and animals. But I had never studied bacterial viruses, or even thought much about them. If I wanted to join Jill on her quest, that had to change.

Frederick Twort, a British bacteriologist working in the early twentieth century, was the first scientist to document the effects of bacterial viruses. Ironically, though, the viruses he initially set out to investigate infected animals and plants — not bacteria — and had already been discovered well before his time. But while Twort was attempting to cultivate these viruses from materials such as dung and hay, he came across a peculiar culture of a bacterium in the genus *Micrococcus*. The sample appeared to be diseased; instead of growing in dense colonies on the nutrient-rich petri dishes, as did most other bacteria, these cultures looked watery and transparent. If he dabbed a bit of the watery-looking *Micrococci* into a healthy culture of *Micrococci*, the healthy culture took on the same glassy appearance, as if it had been infected by something. Twort wrote a paper suggesting that the infectious agent could be a virus, but the idea of a virus infecting bacteria was unheard of at the time, and there were other possible explanations for the transformation. He couldn't say for sure what had sickened the healthy culture.

In 1917, two years after Twort's paper was published, bacterial viruses were rediscovered by a Canadian-born physician named Félix d'Herelle. While stationed in France during World War I, d'Herelle had been assigned to investigate an outbreak of dysentery that was ravaging a cavalry squadron there. Determined to find out why some patients recovered and others didn't, d'Herelle took fecal samples from some of the sick men and subjected them to a rigorous, if crude, analysis. First, he passed the patients' bloody stools through a fine-sieved filter so as to remove all solids — including any bacteria — from the samples. He then spread the filtered liquid on top of cultures of the dysentery-causing *Shigella* bacteria. The next day, d'Herelle was shocked to find that the infectious bacterial culture under one of the liquid samples had "dissolved away like sugar in water," vanishing overnight. Even more remarkable, when he rushed to the hospital to learn the fate of the patient who had donated this particular stool sample, he found the man's condition markedly improved. Putting the pieces together, d'Herelle concluded that a parasite — what he called a *bacteriophage,* or "eater of bacteria," a life form small enough to pass through the filter — had destroyed the *Shigella* bacteria. This bacteriophage seemed to infect bacteria in much the same way that viruses infected plants and animals.

More bacteriophages, or phages (pronounced "*fay*-jehs") for short, were discovered in the years following d'Herelle's experiment, and each was found to target a particular species of bacteria. As the varieties of known phages mounted, excitement grew around what came to be called phage therapy: the notion that bacteriophages could be used to treat bacterial infections. Although some scientists were uncomfortable with the idea of injecting live viruses into human patients, the fact was that bacteriophages appeared to ignore human cells altogether, and patients showed no adverse effects when the therapy was tested in clinical trials. In 1923, d'Herelle helped Soviet scientists set up an institute in Tbilisi, present-day Georgia, dedicated to bacteriophage research; at its peak, the institute had over a thousand employees producing tons of phages a year for clinical use. Phage therapy has continued up to modern times in certain parts of the world — about 20 percent of bacterial infections are treated with phages in Georgia today — but after antibiotics were discovered and developed in the 1930s and 1940s, this treatment quickly lost momentum, especially in the West.

Bacteriophages may have been of limited use as a therapy, but they were a godsend for genetics research. By the time researchers got their first glimpses of phages in the 1940s and 1950s, using new high-magnification electron microscopes, these bacterial viruses, together with the bacteria they targeted, had already provided support for the Darwinian theory of natural selection. They had helped prove that DNA, not protein, was the hereditary molecule of the cell. The fact that genetic code exists in triplets, with three letters of DNA specifying each amino acid of a protein, was demonstrated with phages, and phage experiments also helped reveal how genes are turned on and off inside a cell. Even Joshua Lederberg's discovery that viruses could shuttle foreign genes into infected cells — the early inspiration for gene therapy — resulted from the action of a *Salmonella*-specific bacteriophage. In many ways, the foundations of molecular genetics were laid by experiments done with these bacterial viruses.

Phage research also spawned the molecular biology revolution of the 1970s. While researching the immune systems that bacteria use to fend off phage infections, scientists identified a class of enzymes called restriction endonucleases that could be engineered to cut apart synthetic DNA fragments in simple test-tube experiments. By combining these enzymes with other enzymes isolated from phage-infected cells, scientists were able to design and clone artificial DNA molecules in the lab. At the same time, phage genomes were providing an excellent target for newly invented DNA-sequencing technologies. In 1977, Fred Sanger and his colleagues succeeded in determining the complete DNA genome sequence of a phage called ΦX174. Twenty-five years later, the same phage would again become famous: its genome was the first to be synthesized entirely from scratch.

Bacteriophages aren't just popular laboratory pets, though; they are also the most prevalent biological entity on our planet — by a long shot. They are as ubiquitous in the natural world as light and soil, and they can be found in dirt, water, our intestines, hot springs, ice cores, and

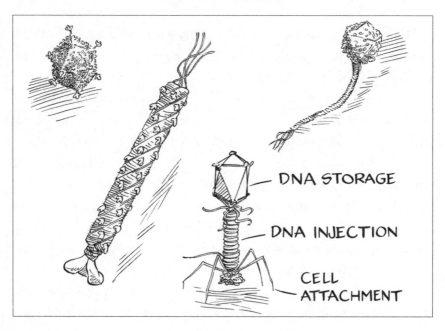

DNA STORAGE

DNA INJECTION

CELL
ATTACHMENT

Examples of different bacteriophages

just about anywhere else that supports life. Scientists estimate that there are somewhere on the order of 10^{31} bacteriophages on earth; that's ten million trillion trillion, or a one with thirty-one zeros after it. A single teaspoon of seawater contains five times more phages than there are people in New York City. Incredibly, there are many, many more phages on earth than there are bacteria for them to infect; as abundant as bacteria are, bacterial viruses outnumber them ten to one. They cause roughly a trillion trillion infections on earth every second, and in the ocean alone, about 40 percent of all bacteria die every day as a result of deadly phage infections.

These viruses are lethal by design, having evolved over billions of years to infect bacteria with brutal efficiency. All phages are built out of a durable protein exterior, called a capsid, inside of which genetic material is packaged. Phage capsids come in dozens of different shapes, all of which have been optimized to safeguard the viral genome and effectively deliver the genetic material into bacterial cells, where it can multiply and

spread. Some phages have elegant icosahedral (twenty-sided) geome-
tries; others have spherical capsids attached to long tails. Filamentous
phages are cylindrical. Perhaps the scariest of these viruses is the one
that looks like an alien spacecraft, with legs to latch onto the outer surface of the cell, a head where the DNA is stored, and pumps that inject
that DNA into the cell after the phage has landed.

The viruses' modi operandi, like their appearances, are diverse but
invariably, and ruthlessly, effective. Some viral genomes are packed so
tight in the capsids that the genetic material explodes into the cell as
soon as the protein shell is breached, releasing internal pressure like an
uncorked bottle of champagne. Once the genome makes its way inside
the cell, it can hijack the host by one of two possible pathways. In the
parasitic, or lysogenic, pathway, the viral genome insinuates itself into
that of its host, where it can stay buried for many generations, waiting for
the right moment to strike. By contrast, in the infectious, or lytic, path-

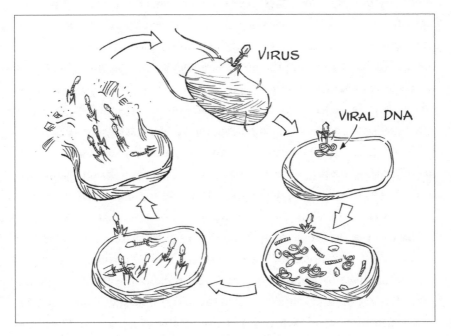

The life cycle of a bacteriophage

way, the genome commandeers its host's resources immediately, direct-ing the bacterium to produce viral proteins instead of bacterial proteins and replicate the viral genome many times over until the cell violently bursts open from the mounting pressure and scatters fresh phages that infect neighboring cells. Through this cycle of cellular invasion, hijack-ing, replication, and propagation, a single phage can wipe out an entire bacterial population in a matter of hours.

But bacteria aren't powerless in this age-old war. Just like plants and animals, they've developed impressive defense strategies over billions of years of evolution. At the time I spoke with Jill, four major bacte-rial defense systems had been identified. In the most prominent one, bacteria decorate their own genomes with unique markings that subtly change the DNA's chemical appearance without affecting how genetic information is expressed; in addition, bacteria unleash enzymes called restriction endonucleases to chop up any DNA that lack those markings, effectively purging any phage genes that managed to penetrate the cell wall. Bacteria can also block phage DNA from ever making it inside the cell, either by plugging holes made by phages so that they can't inject their DNA or by masking the protein molecules on the outer surface of the cell that phages latch onto. Bacterial cells have even developed meth-ods to sense an oncoming infection and commit suicide before it can progress — a selfless way of protecting the greater bacterial community.

Could CRISPR be yet another antiviral defense mechanism? The more I read about the arms race being waged between bacteria and bac-teriophages, the more excited I got about the possibility that there was another weapons system out there, waiting to be discovered.

What's more, as I read up on CRISPR, I began to gain a sense of where I would direct my lab's efforts if we did take up Jill's challenge. Compu-tational analyses by Ruud Jansen and his colleagues in the Netherlands — the same team that initially coined the CRISPR acronym back in 2002 — had identified a set of genes that almost always flanked the CRISPR regions in bacterial chromosomes. These weren't the repeat sequences or the spacer sequences within the CRISPR DNA but a separate set of genes entirely.

From what little we knew about them, these CRISPR-associated genes, or *cas* genes, seemed full of exciting potential. Comparisons with known genes suggested that *cas* genes coded for specialized enzymes whose functions might include unzipping the two strands of the DNA double helix or slicing up RNA or DNA molecules, just like the DNA-cutting function of restriction endonucleases.

Given how useful the discovery of restriction endonucleases had been for recombinant DNA technology in the 1970s, it seemed very possible that, by digging deeper into these and other aspects of CRISPR, we might uncover a treasure-trove of new enzymes — and that these proteins too could have major biotechnology potential.

That was it. I was hooked.

Research scientists are fueled by adventurousness, curiosity, instinct, and grit — but we need a healthy sense of practicality in addition to these loftier traits. There are the banalities of funding to consider and management requirements galore. Those of us who run our own labs need to delegate to other scientists many of the tasks we were trained to do ourselves. Often, this means selecting the right person to lead the charge whenever we enter a brand-new field of research.

I was lucky enough to have considerable financial support for my lab at Berkeley, but when Jill first introduced me to CRISPR, there wasn't anybody on my team who was qualified to take on this unpredictable and potentially risky new project. Then, as luck would have it, Blake Wiedenheft showed up to interview for a position as a postdoctoral scientist in my lab. When I asked the young applicant what he wanted to work on, I was delighted to hear him answer with a question of his own: Have you ever heard of CRISPR? I hired him on the spot. Just a few months later, Blake was comfortably situated in Berkeley, working furiously to get our CRISPR project off the ground.

A warm and engaging Montana native with a competitive streak born of his love for outdoor sports, Blake came to Berkeley from Bozeman, where he had completed both his undergraduate and graduate degrees at Montana State University. Unlike most of the scientists I had hired before him, people with expertise in biochemistry or structural biology, Blake was a hard-core microbiologist. Like Jill, he'd spent part of his career in the lab and part of it out in the field collecting samples. His PhD research had taken him to both Yellowstone National Park and Kamchatka, Russia, where he had discovered novel viruses lurking in the acidic waters of hot springs, intact and infectious despite temperatures that rose to over 170 degrees Fahrenheit. These viruses were known to infect archaea — those single-celled microorganisms similar to bacteria and in whose genomes CRISPR was almost ubiquitous. After sequencing the genomes of two viruses he had isolated, Blake found they shared extensive amounts of DNA; this meant that, although a huge geographic distance separated Yellowstone and Kamchatka, they had to share a common ancestor. The genomes also contained clues about how the viruses infected their hosts; by analyzing specific viral genes, Blake singled out one enzyme that, he suspected, allowed the viruses to splice their genomes into the DNA of their unsuspecting hosts.

It was exactly this kind of detective work that we needed to apply to CRISPR, only in reverse. Instead of focusing on viral genes that promoted infection, we had to hunt down the genes in bacteria that blocked infection — the ones associated with CRISPR. Or, rather, the bacterial genes that we *thought* blocked infection. We still weren't sure that's what *cas* genes, or CRISPR itself, actually did.

Much of our early brainstorming centered on this alluring hypothesis: that CRISPRs and *cas* genes were parts of the same antiviral immune system and that RNA was employed by this system to detect viruses. But a hypothesis is only the first step in any rigorous scientific process. We still needed to test it and gather evidence that either supported or refuted our theory.

In meetings with Jill and a handful of interested scientists at the Lawrence Berkeley National Laboratory, just a short walk from my office, Blake and I deliberated over how to set up the experiments. A big question was what model organism we should use. One option was *Sulfolobus solfataricus*, an archaeal microorganism first isolated from hot springs in the Solfatara volcano, near Naples, Italy. This archaeon was known to contain CRISPR, and it could also be infected by Blake's Yellowstone and Kamchatka viruses, which was convenient because Blake knew them so well. Another option was *Escherichia coli*, commonly known as *E. coli*. By far the most well-studied bacterial species in microbiology, *E. coli* is susceptible to infection by dozens of equally well-studied phages, many of which could simply be purchased online. (*E. coli* also had the distinction of being the first bacterium in which a CRISPR sequence had been identified.) In addition, Blake proposed *Pseudomonas aeruginosa*, a disease-causing bacterium known to be resistant to many antibiotics and one that possessed a CRISPR. We knew that we'd be able to manipulate *P. aeruginosa* using genetic tools and also that it could be infected by numerous phages. (Blake would spend some time hunting for new *Pseudomonas* phages, not in exotic locales like Yellowstone, but in local Bay Area sewage treatment plants.)

Blake was very clear with me that he wanted to focus on learning biochemistry and structural biology during his stay in my lab, and he was eager to strike out in a new scientific direction. To begin work on CRISPR, he purified the Cas proteins that were encoded in the *P. aeruginosa* genome and began testing them for the ability to somehow recognize or destroy viral DNA, starting with the most widely occurring Cas protein, Cas1. Then in 2007, around the same time that Blake started in my lab, we got word from Jill about an exciting paper soon to be published by scientists at Danisco, a Danish bio-company and one of the world's leading food-ingredient producers. Their study showed, using genetics, that CRISPR was indeed a bacterial immune system — although the details of its capabilities were still unknown.

The subject of the Danisco study was a milk-fermenting bacterium called *Streptococcus thermophilus,* one of the key probiotics involved in producing yogurt, mozzarella cheese, and numerous other dairy products. Humans ingest well over a billion trillion live *S. thermophilus* cells a year, and the annual market value of cultures of the bacterium is more than forty billion dollars. Perhaps unsurprisingly, this massive investment by the dairy industry is under constant threat from phage infection, the most common cause of production losses and incomplete fermentation. A single drop of raw milk contains anywhere from ten to one thousand virus particles, making phage eradication simply impossible. Companies like Danisco had tried to combat phages with improved sanitation, new factory designs, and other tactics — but nothing had solved the problem.

Working with Philippe Horvath and his team at Danisco France, a group of researchers led by Rodolphe Barrangou at Danisco USA had been studying *S. thermophilus* to see if they could find a different solution. Rodolphe and Philippe wondered what made some strains of *S. thermophilus* more resistant to phage infection than others. The dairy industry had already deployed some mutant bacterial strains that were less susceptible to bacteriophages, but Rodolphe and Philippe suspected that the CRISPR regions of the *S. thermophilus* genome might provide a form of immunity that could be even more powerful than these random mutations.

The *S. thermophilus* CRISPR sequences, Rodolphe and Philippe knew, had some intriguing characteristics that they could exploit for their research. A scientist named Alexander Bolotin had uncovered some of these characteristics when he sequenced the bacterium's genome and later focused on the CRISPR DNA in particular, eventually expanding his analysis to include more than twenty different strains. In the process, he'd observed that, although the repeat sequences of CRISPR (the shaded black diamonds in Jill's sketch) were always the same, the spacer sequences (Jill's numbered squares) were highly variable from one strain to the next. Furthermore, many of these spacers perfectly matched parts of phage genomes that had recently been sequenced. (Bolotin's findings had been summarized in one of the three papers from 2005 that Jill showed me at the Free Speech Movement Café.) The kicker from Bolotin's publication: strains of S. thermophilus that had more of these spacers seemed to be more resistant to phage infection. While it wasn't clear what this meant, it appeared that the bacteria had somehow modified their CRISPR DNA to mirror certain phage genomes, upgrading their own immune system — assuming that's what CRISPR was — to more effectively fight off those viruses.

Building on Bolotin's work, Rodolphe and Philippe designed experiments to test this hypothesis. Could a strain of S. thermophilus actually make itself more resistant to a particular bacteriophage by splicing new DNA into its own CRISPR region that matched the sequences of DNA found in the phage?

For their experiments, the Danisco researchers focused on a strain of S. thermophilus that was widely used in the dairy industry and on two virulent phages that had been isolated from industrial yogurt samples. Taking their cue from one of the simplest genetics experiments, the kind that had been conducted since the early twentieth century, they combined the bacterial strain with the two phages in separate test tubes, let them incubate for twenty-four hours, then checked to see whether any bacteria were still alive by spreading the cultures onto petri dishes and letting them grow overnight. They found that, although the phages wiped out well over 99.9 percent of the bacteria, nine new mutant strains of S. thermophilus seemed to be impervious to the phages.

Up until this point, nothing about the Danisco experiment had been particularly novel, since other scientists had used similar methods to isolate phage-resistant strains of S. thermophilus. But Rodolphe and Philippe took their investigation further. They attempted to pinpoint the genetic cause of this apparent immunity.

Rodolphe and Philippe had a hunch about which part of the bacterial

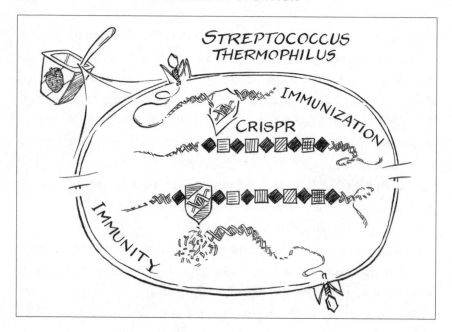

CRISPR: a molecular vaccination card

genome had made these mutant strains of *S. thermophilus* immune to viruses. They suspected it was CRISPR and hypothesized that the CRISPR regions of these nine new mutant strains would look different than the CRISPR of the original strain. Sure enough, after isolating the genomic DNA from each mutant strain, the researchers found that every single CRISPR region had expanded to include a new snippet of DNA spliced between the repeats. Furthermore, these new spacers perfectly matched the DNA of the phage to which that strain was now immune. What made this apparent mode of immunity so remarkable was that, because these changes were physically embedded in the bacteria's CRISPR DNA, the new immunity was heritable and would be passed down every time the bacterial cells reproduced.

The Danisco researchers had revealed another way bacteria fought viruses — a fifth weapons system. In addition to their previously discovered defenses, we now knew, bacteria had in CRISPR a remarkably effective form of adaptive immunity, one that allowed the bacterial genome to steal snippets of phage DNA during an infection and use it to mount a future immune response. As Blake put it, CRISPR functioned like a molecular vaccination card: by storing memories of past phage infections in the form of spacer DNA sequences buried within the repeat-spacer arrays, bacteria could use this information to recognize and destroy those same invading phages during future infections.

Publication of the Danisco study began to attract attention to the obscure biology of CRISPRs, and it also catalyzed the first CRISPR meeting, held at UC Berkeley and organized by Jill Banfield and Rodolphe Barrangou in 2008. However, as is always the case in science, researchers had broken down one door only to be confronted with another. Since the CRISPR immune response required DNA sequences in the bacterial genome and the phage genome to match up perfectly, it was clear that this immune system was targeting the phages' genetic material for destruction — but how? What part of the cell was doing the targeting?

It wasn't long before an answer to this new question began taking shape. Stan Brouns, a postdoctoral scientist working in John van der Oost's lab at the University of Wageningen in the Netherlands, soon provided unambiguous evidence that molecules of RNA were involved in CRISPR's antiviral defense. Stan had been building on earlier research that had detected RNA molecules that precisely matched the sequence of CRISPR DNA within the cells of various archaeal species, including the volcanic *Sulfolobus* strains that Blake had studied. This led to speculation that RNA might coordinate the recognition and destruction phases of bacteria's antiviral response. Now Stan, experimenting with *E. coli*, had advanced these observations by confirming that RNA played this role in the CRISPR defense system of a completely different kind of microorganism — good evidence that RNA was universally required for CRISPR-related immune systems.

Stan also showed how CRISPR RNA molecules were produced inside the cell. First, the bacterial cell converted the entire CRISPR array into long RNA strands that matched the sequence of CRISPR DNA exactly, letter by letter. (Remember that RNA is a molecular cousin of DNA, made up of virtually the same letters, except the letter T in DNA is replaced with the letter U in RNA.) Once the cell had created these long strands of CRISPR-derived RNA, an enzyme surgically trimmed them to shorter RNA strands of uniform length, the only difference between them being the sequence of their spacers. This process converted the long repetitive arrays in DNA into a library of shorter RNA molecules, each containing a single sequence derived from a particular phage.

These findings hinted at the vital role CRISPR RNA plays within the bacterial immune system — a role made possible by the basic functions of RNA itself. Since RNA is chemically so similar to DNA, it can create double helixes of its own by using base-pairing interactions, the same process that forms the famous double helix of DNA. Matching RNA strands can pair with each other, forming an RNA-RNA double helix, but a single strand of RNA can also pair with a matching single strand of DNA, forming an RNA-DNA double helix. This versatility, and the variety of different sequences found in CRISPR RNA, gave scientists an intriguing idea. It seemed possible that these CRISPR RNA molecules could single out both DNA and RNA molecules from invading phages for attack during an infection by pairing with any that they matched and initiating some sort of immune response in the cell.

If RNA did help to target viral genetic material in this way, then CRISPR might indeed be similar to the RNA interference pathway that my lab was studying — just as had been posited in that paper that lured me into CRISPR research in the first place! In RNA interference, animal and plant cells form RNA-RNA double helixes to destroy invading viruses. In much the same way, CRISPR RNA molecules might target phage RNA during an immune response by using RNA-RNA double helixes. I was fascinated by the added possibility that, unlike in RNA interference, CRISPR RNAs might be able to recognize matching DNA too — a power that would enable this weapons system to attack the viral genome on both fronts.

Shortly after Stan's discovery, two researchers at Northwestern University, Luciano Marraffini and his mentor Erik Sontheimer, a colleague I knew from his student days at Yale, figured out that CRISPR RNA could, indeed, direct the destruction of DNA. Working with yet another microorganism called *Staphylococcus epidermidis,* a relatively benign human skin bacterium (but a close relative of the dangerous, drug-resistant strain of *Staphylococcus aureus*), Luciano used a series of elegant experiments to prove that CRISPR RNAs target the DNA of invading genetic parasites. He also showed that this targeting likely relied on base-pairing interactions — the only process that could account for the specificity with which CRISPR hunted its prey.

The pace and rigor of these studies was breathtaking. Within just a few years of my introduction to CRISPR, the field had advanced from a loose collection of interesting but inconclusive studies to a broad, unified theory about the inner workings of a microbial adaptive immune system. This theory was based on a growing body of experimental research, but while many landmark studies had been published by the late 2000s, it was clear that we had much more work to do before we could truly comprehend this intricate bacterial defense system.

CRISPR, we were beginning to understand, was far more complex than anyone had imagined possible for simple, single-celled organisms. In some respects, the discovery of this part of the bacterial immune system placed bacteria on an equal footing with humans by showing that we both have incredibly complex cellular reactions to infection. Just what implications these bacterial defenses might have for our species, however, none of us knew.

3

CRACKING THE CODE

I REMEMBER THE FIRST TIME I stepped into a professional research laboratory — the sounds, the smells, and the sense of possibility, of nature's secrets being slowly uncovered. It was 1982 and I was back with my parents in Hawaii after my freshman year in college. My father, a professor of English at the University of Hawaii, had arranged for me to spend a few weeks in the lab of a colleague of his, biology professor Don Hemmes. Along with two other students, I would be investigating how a fungus, *Phytophthora palmivora*, infected papayas. A big problem for fruit growers, the fungus turned out to be a lot of fun to study. It grew quickly and easily in the lab and could be trapped in different states of germination, allowing us to detect chemical changes that took place as it developed. That summer I learned how to embed fungus samples in resin and shave off thin sections for analysis by electron microscopy. While my time on the project was brief, our work revealed something important about the fungus: calcium ions play a crucial role in its development by signaling the fungal cells to grow in response to nutrients. It was my first taste of the thrill of scientific discovery, an experience that I'd read so much about — and it left me hungering for more.

The peace and quiet concentration that characterized Don Hemmes's small research team drew me in, but over the years I became aware of being part of a much bigger community of scientists, each of us seeking, in our own ways, nature's truths. With every tiny advance, it felt like we were finding another piece in an enormous jigsaw puzzle, one in which each person's work built on the work of others to fill in more of the picture.

The CRISPR project epitomized this aspect of science: a few researchers around the world weaving the fabric of what would eventually become the vast tapestry of the CRISPR field, with all of its applications and implications. And in the quest to learn more about CRISPR, our little team and many others were driven by the same sense of collaboration, of shared excitement and curiosity, that had pulled me into the world of scientific research in the first place.

In those early days of the field, Blake and I were energized by the work of our colleagues at Danisco, Northwestern, and Wageningen, and at the same time intrigued by the fact that many fundamental questions about CRISPR remained unanswered. Although biologists now appreciated that CRISPRs provided bacteria and archaea with adaptive immunity against phages and that phage DNA sequences matching CRISPR RNA were somehow targeted for destruction, nobody really knew *how* this all worked. We wondered how the various molecules that made up such a system acted together to destroy viral DNA and what exactly happened during the targeting and destruction phases of the immune response.

As these questions came into sharper focus, so did the challenges. We needed to find out how bacteria could pilfer short segments of DNA from a phage genome in the midst of an infection and precisely integrate these segments into the existing CRISPR array so that the defense system could target the virus's genetic material. We needed to determine how CRISPR RNA molecules were produced inside the cell and converted from long strands into much shorter pieces, each containing a single virus-matching sequence. And perhaps most important, we had to find out how a piece of RNA could pair up with its phage DNA counterpart and cause that DNA to be destroyed. That was the crux of this new weapons system, and we wouldn't fully understand CRISPR until we understood this part of the process.

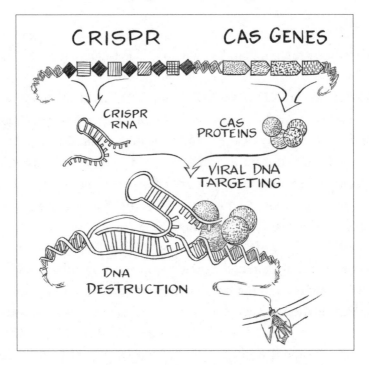

DNA targeting by CRISPR RNA and Cas proteins

It was clear that addressing these questions would require us to move beyond genetics research and take a more biochemical approach — one that would allow us to isolate the component molecules and study their behavior. We would also need to broaden our attention from CRISPR itself to all the CRISPR-associated genes, or *cas* genes, that flanked the CRISPR regions of bacterial genomes and that appeared to contain the code for special types of proteins called enzymes. Generally speaking, this class of protein molecules is responsible for catalyzing all sorts of molecular reactions within cells. Odds were that, if we could find out what those Cas protein enzymes did, we'd be a lot closer to understanding how CRISPR worked.

Scientists can learn a lot about a gene's function just by looking at its chemical composition. The stretches of DNA that make up each gene contain all the information a cell needs to assemble proteins from amino acids. Because we know the genetic code that cells use to translate the four letters of DNA into the twenty letters of protein, biologists can determine the amino acid sequence of the protein that a gene will produce just by looking at the original DNA sequence. Then, by comparing that amino acid sequence to other, related proteins that are better understood, scientists can make informed predictions about the functions of many different genes.

Using this kind of sophisticated guesswork, computational biologists had already figured out the chemical makeup of hundreds of different *cas* genes that invariably coexisted with CRISPR regions. No matter which organism you were interested in, if its genome contained CRISPR DNA, there were bound to be *cas* genes in the immediate vicinity. It was almost as if CRISPR had co-evolved with *cas* genes; it didn't seem possible to have one without the other.

The proteins encoded by these *cas* genes, we reasoned, must work intimately with CRISPR DNA — or perhaps CRISPR RNA molecules, or even phage DNA. One thing seemed certain: we'd need to find out how these genes worked and discern the biochemical functions of the proteins they produced before we could understand the CRISPR immune system as a whole.

To start, Blake picked two bacterial species, *Escherichia coli* and *Pseudomonas aeruginosa,* that contained distinct types of CRISPR systems. The *E. coli* bacterium, in particular, is a biochemist's best friend. Whether she or he is studying a gene from a microbe, a plant, a frog, or a human, a biochemist will often begin by cloning that gene into an artificial mini-chromosome, called a plasmid, and then engineering a specialized strain of *E. coli* to accept that plasmid as part of its own genome. By piecing together the gene of interest with other synthetic DNA instructions, the biochemist can trick *E. coli* into not only churning out dozens of copies of that plasmid per cell but also dedicating the majority of its resources to converting the gene of interest into thousands of copies of the protein that it encodes. In this way, the biochemist reduces *E. coli* to little more than a microscopic industrial bioreactor, programmed to produce specific proteins on a massive scale.

Blake quickly built plasmids out of individual CRISPR-associated genes he had copied from the genomes of both *E. coli* and *P. aeruginosa.* Amassing dozens of *E. coli* strains that he had engineered to incorporate these plasmids as part of their own genetic material, Blake began growing up liters of cultures of each of the engineered strains to produce enough of the Cas proteins for his experiments. After allowing the bacteria to grow overnight, Blake would dump the contents of the culture flasks into large bottles and separate the cells from the liquid broth in large, fast-spinning centrifuges, where they were subjected to forces four thousand times greater than the gravity we feel on earth. Next, working separately with each strain, he would suspend the cells in a small volume of salt solution and subject this slurry to high-energy sound waves that violently burst open the cells, liberating their contents — including the Cas proteins they had produced.

After discarding the debris — ruptured membranes, viscous DNA, and other types of cellular gunk — Blake was left with the few thousand proteins in the cell, of which he wanted only one: the Cas protein. But thanks to the ingenious design of the plasmid, the Cas protein contained a special chemical tag, or appendage, that distinguished it from all those other thousands of proteins. Using a purification strategy to weed out this molecular appendage, followed by a series of additional refinements, Blake could obtain pure, highly concentrated samples of each of the Cas proteins we were interested in studying.

With the Cas proteins in hand, Blake could finally go about setting up different kinds of experiments to investigate what these enzymes did. In our first contribution to the CRISPR field, we published our discovery that a protein enzyme called Cas1 had the ability to cut up DNA in a way that suggested it might help insert new snippets of phage DNA into the CRISPR array during the immune system's memory-forming stage. This got us one step closer to understanding how CRISPR stole bits of DNA from attacking phages and worked that genetic information into its own, laying the groundwork for the targeting and destruction phases of the immune response.

Around that time, Blake recruited a new graduate student, Rachel Haurwitz, to the CRISPR project, and together they made another discovery. Working with a second protein enzyme, Cas6, Rachel and Blake found that, like Cas1, it functioned as a chemical cleaver. In the case of Cas6, however, its job was to specifically and methodically slice the long CRISPR RNA molecules into shorter chunks that could be used to target phage DNA.

As we and others amassed these pieces of the CRISPR puzzle, slowly but surely an image began to take shape. In it, we could already discern the answers to some of the questions we had posed at the outset of the project. And there seemed to be no shortage of Cas protein functions to uncover. In the course of our research, we found more and more Cas proteins that were DNA- or RNA-cutting enzymes and that therefore seemed to play roles in the CRISPR immune response similar to those of Cas1 and Cas6.

By 2010, the CRISPR project had expanded to include several other members of my team, including coauthor Sam Sternberg, and the atmosphere of the lab was electric with excitement. Our understanding of CRISPR seemed to grow every week or two, and the enzymes we were exploring had many interesting, unusual properties — properties that, we recognized, might have practical uses. For example, we started toying around with the idea of developing the new RNA-cutting enzymes into a kind of diagnostic tool to detect signature RNA molecules in human viruses, including dengue virus and the yellow fever virus; we received money from the Gates Foundation to put that idea into practice. Soon we partnered with a bioengineering lab at Berkeley to combine this technology with their innovative system for manipulating tiny amounts of liquid to detect viruses in blood or saliva.

Then in 2011, Rachel and I founded a company called Caribou Biosciences to commercialize the Cas proteins. At the time, we imagined creating simple kits that scientists, or even clinicians, could use to detect the presence of viral or bacterial RNA in body fluids. For both Rachel and me, this departure from the academic world took us into an exciting new realm. After completing her PhD the following spring, Rachel became the fledgling company's president and CEO; I was a scientific adviser — a role that enabled me to contribute to Caribou's endeavors while keeping up with my duties on campus. Eventually, Caribou would become famous for another, much more powerful, CRISPR-related technology.

During this time, Blake's focus and my interests shifted away from the enzymes that were involved in cutting bacterial CRISPR DNA or RNA molecules and toward those proteins that had the demanding job of cutting up viral DNA — the task that constituted the destruction phase of CRISPR's search-and-destroy process. Once CRISPR RNA had identified and paired with viral DNA, we imagined, special enzymes attacked this foreign genetic material, chopping it into pieces and disabling it. Interesting evidence in support of such hypotheses was coming from research conducted by other colleagues in the field, including Sylvain Moineau at Laval University in Canada and Virginijus Siksnys at Vilnius University in Lithuania. Sylvain's research showed that phage DNA targeted by the CRISPR system got sliced apart within the sequence matching the CRISPR RNA, and Virginijus found that phage eradication in bacteria depended on the presence of specific *cas* genes. Figuring out how a phage's genetic material ultimately got destroyed during an immune response would really get to the heart of the entire CRISPR pathway.

Blake's research, in conjunction with that of our collaborators in John van der Oost's lab, began to reveal just how complicated this process of killing a virus was. In *E. coli* and *P. aeruginosa*, the two bacterial systems we were studying, cells required multiple Cas proteins to target and cleave viral DNA. In addition, the coordinated attack on the phage's genetic material proceeded in two distinct phases. First, the CRISPR RNA molecule got loaded into a much larger assemblage, containing some ten or eleven different Cas proteins, as the van der Oost lab had shown. This molecular machine — John's lab had colorfully dubbed it Cascade (yet another biologist's acronym, standing for "CRISPR-associated complex for antiviral defense") — acted like GPS coordinates, defining the exact sequence of viral DNA to be destroyed. In the second phase, after Cascade had located and marked the matching DNA sequence for destruction, a protein enzyme called Cas3 — another nuclease, and the actual weapon in the attack — swooped in to cut apart the targeted DNA.

As we performed the experiments for a series of articles we published in 2011 and 2012, the mechanics of this process became even clearer. Using the powerful beams of an electron microscope, and working closely with Berkeley professor Eva Nogales and her postdoctoral associate Gabe Lander, we obtained the first high-resolution images of the Cascade machine. These pictures revealed the helical architecture of these Cas proteins and CRISPR RNA molecules and showed how the microscopic machine snugly wrapped around viral DNA, like a python coiling around a gazelle. We saw, thrillingly, how its three-dimensional shape had beautifully evolved to match the geometric needs of its DNA targeting function. We also discovered the importance of base-pairing interactions that allowed the letters of CRISPR RNA to recognize the matching letters of viral DNA and found that Cascade was remarkably adept at locking onto only those viral DNA targets that were perfect or near-perfect matches to the CRISPR RNA. This high degree of discrimination allowed Cascade to avoid accidentally targeting the bacterium's own DNA for destruction, a catastrophic autoimmune event that would rapidly trigger cell death.

Complementary studies from Virginijus Siksnys's lab in Lithuania showed how the Cas3 enzyme destroyed the viral DNA targeted by Cascade. Unlike simpler nucleases, Cas3 didn't cut the DNA just once; it chewed it up into hundreds of pieces. Once Cascade recruited Cas3 to the site of a CRISPR RNA–viral DNA match, Cas3 started shuttling along the phage genome at a rate of over three hundred base pairs per second, slicing up the DNA and leaving the lengthy phage genome as a jumble of scraps in its wake. If simpler nucleases were like pruning shears, Cas3 was like a pair of motorized hedge clippers. Its speed and efficiency were stunning.

As our fellow researchers produced fascinating findings like this, and as my lab continued its contributions of biochemical and structural data, the previously hazy inner workings of CRISPR began to coalesce into a discrete set of molecules carrying out discrete actions. At the same time, however, we were finding that the CRISPR immune system was a bit of a moving target; instead of a single CRISPR immune system, there appeared to be many variations — something that scientists, including Eugene Koonin and Kira Makarova, had predicted based on comparisons of the different sets of *cas* genes found flanking the CRISPR arrays. We were discovering this thanks to a massive increase in the number of bacterial and archaeal genomes being sequenced by researchers with easier access to better sequencing tools. CRISPR immune systems were turning out to be highly diverse and could be grouped into multiple different categories, each with its own unique complement of *cas* genes and Cas proteins.

We were amazed to see just how diverse CRISPR was. In 2005, researchers had identified nine different types of CRISPR immune systems. By 2011, that number had decreased to three — but within these basic types there were thought to be ten subtypes. And by 2015, the classification would change yet again to include two broad classes comprising six types and nineteen subtypes.

These findings put our own research into perspective — and made clear the limitations of our work thus far. The results we had been assembling for *E. coli* and *P. aeruginosa* were true of only two of those subtypes of CRISPR systems, which were themselves part of what was known as the Type I CRISPR-Cas immune system. While many of the conclusions from our research applied to other subtypes of CRISPR, it was becoming increasingly difficult to compare our data to bacteria with Type II systems like *S. thermophilus,* that yogurt-producing bacteria in which CRISPR-based immunity had first been demonstrated.

There were also some bizarre differences in the way that the various CRISPR-Cas systems destroyed phage DNA. In Type I systems like *E. coli* and *P. aeruginosa,* the Cas3 enzyme — that motorized hedge clipper — chewed the DNA to shreds. It wasn't even possible to see the DNA destruction in action because this tiny machine mowed through it so rapidly; when we tried to observe the reaction in a test-tube experiment, all we could see was molecular chaos, with big long smears of DNA obliterated all along the phage's genome. The Type II system found in *S. thermophilus,* by contrast, was more restrained and precise. Canadian scientists Sylvain Moineau and Josiane Garneau, working with the Danisco team, had succeeded in trapping phage genomes from infected cells as they were being destroyed by the CRISPR immune system. In a process typical of simpler nucleases, whatever was doing the cutting in *S. thermophilus* operated more like a pair of scissors, clipping the DNA apart at exactly the site where the letters of the viral genome matched the letters of the CRISPR RNA.

The surgical precision of the Cas enzyme in *S. thermophilus* was impressive — but we didn't know nearly as much about the protein in this Type II system as we did about the enzyme in the Type I system. None of the proteins in the Cascade machine Blake and I had studied, which was solely responsible for targeting DNA in the Type I CRISPR system of *E. coli,* was even present in the Type II system of *S. thermophilus.* What's more, we weren't sure how the Type II enzymes worked together with the CRISPR RNA to specify where along the viral DNA to cut.

What enzyme was the tip of the spear in the Type II system if not Cas3? What, for that matter, was doing the DNA targeting in this system — CRISPR RNA together with which other actors? Answering these questions would allow us to understand how nature had solved this same molecular challenge — destroying viral DNA — in distinct ways, and it would also help us to comprehend, and perhaps even tame, a powerful new type of bacterial immune system.

This mysterious defense system created by nature had features oddly reminiscent of engineered nucleases, the programmable DNA-cutting enzymes that were increasingly being deployed to induce precise DNA changes in cells in the process known as gene editing. Although the Type II CRISPR immune system in *S. thermophilus* appeared to destroy phage DNA rather than edit it, its ability to find and cut specific DNA sequences was no different, at least in principle, from the functions of gene-editing tools that were already in existence: the ZFNs and the TALENs. But there were also important differences, and two questions in particular were drawing the attention of CRISPR researchers: What kind of enzyme was doing the DNA cutting in this Type II system, and how did it work?

Since I was still focused on Type I systems, I might not have been drawn into tackling this new line of research had it not been for a fortu- itous encounter with a colleague whose lab was located halfway across the globe and whose work I had only read about. Our chance meeting would inspire me to focus our efforts to understand CRISPR in a new direction and would turn into a life-changing collaboration — one that revealed a side to this wondrous system few could have imagined.

In the spring of 2011, I traveled from Berkeley to Puerto Rico to attend the annual meeting of the American Society for Microbiology. Confer- ences like these are a great way for scientists to meet new colleagues, catch up on new developments in their particular field, and take a break from the daily grind of life in the lab. Although the microbiology meet- ing was not one that I regularly attended, I had been invited to present my lab's work on CRISPR, and I knew that John van der Oost, by now a friend and occasional collaborator, would also be there. I was excited to talk to him in person and also to explore Puerto Rico. I had visited the island once before, while in graduate school, and I recalled how much its lovely rainforests and ocean vistas reminded me of my hometown in Hawaii.

On the evening of the second day of the conference, John and I popped into a café to load up on coffee before heading into the auditorium where the next session of presentations would take place. Sitting in the corner of the café was a stylish young woman. John brought me over and introduced us, and as soon as he mentioned her name — Emmanuelle Charpentier — a light bulb went on in my head.

The students in my lab had told me about the fascinating talk Emmanuelle had given at a small CRISPR meeting in Wageningen the previous year. I hadn't been able to attend, but my lab members who were there had commented on Emmanuelle's presentation about the Type II CRISPR immune system in a bacterium called *Streptococcus pyogenes.* I realized, connecting the dots, that her paper on the same topic had recently been published in the journal *Nature,* and it had caused a flurry of excitement in my lab. Until its release, everyone in our world had thought that there was just one kind of RNA molecule involved in CRISPR pathways. But Emmanuelle and collaborator Jörg Vogel, whose lab had long studied the functions of small bacterial RNAs, had serendipitously discovered a second RNA molecule that was necessary, in some cases, for the production of CRISPR RNAs. This finding had thrilled fellow CRISPR researchers because it showcased the fascinating diversity of bacterial immunity, suggesting that evolution had generated a virtual Swiss army knife to fight viruses.

In our brief conversation, Emmanuelle came across as soft-spoken and retiring but also slyly humorous and refreshingly lighthearted. I took an immediate liking to her. The next day, after the morning talks, we had a free afternoon. I had planned to sit on the patio and do some work on my computer, but Emmanuelle invited me to explore Old San Juan with her, and I couldn't resist. As we strolled through the cobbled streets, which Emmanuelle said reminded her of her childhood home in Paris, we chatted about our recent travels, compared notes about the university systems at Berkeley and in Sweden (where her lab had recently relocated), and reviewed the conference talks we'd heard thus far. Eventually, our conversation shifted to our own scientific research, and Emmanuelle said she had been meaning to call me for some time to propose a collaboration.

Emmanuelle was excited to determine how the Type II CRISPR system in the infectious bacterium she was studying, *Streptococcus pyogenes*, snipped apart viral DNA. Her research, along with earlier genetic studies by Sylvain Moineau, Virginijus Siksnys, and their colleagues, was increasingly implicating at least one gene, called *csn1*, as likely being involved. Would I consider joining forces with her and applying my lab's expertise in biochemistry and structural biology to help figure out the function of the protein encoded by the *csn1* gene? As we walked down a narrow street toward the sparkling blue ocean beyond, Emmanuelle turned to me. "I'm sure that by working together we can figure out the activity of the mysterious Csn1." I felt a shiver of excitement as I contemplated the possibilities of this project.

I was intrigued by the opportunity to do some work with Type II CRISPR systems, those that lacked the Cascade and Cas3 proteins. And if this mysterious Csn1 protein really was involved in DNA cutting in Type II systems, then partnering with Emmanuelle could give my lab a chance to contribute to this area of the CRISPR field.

I was tantalized by this new bacterium too. As a test subject, *S. pyogenes* had some interesting similarities and differences to *S. thermophilus*, the yogurt-culturing bacteria that had by then become one of the preferred organisms in which to study CRISPR. For one thing, both belonged to the same genus, *Streptococcus*, and the CRISPR immune system in *S. pyogenes* looked very similar to that in *S. thermophilus*. Although each bacterium targeted a unique set of phages, they both contained the same core molecular components and all the same genes, which would make it easy to transition from studying one to studying the other.

Yet *S. pyogenes* plays a very different role in our lives than *S. thermophilus* does. Research into *S. thermophilus* has economic value because of the bacteria's widespread use in the dairy industry to produce cheese and yogurt. Notably, *S. thermophilus* is the lone bacterial strain in the *Streptococcus* genus that is generally recognized as safe in humans and other mammals. *S. pyogenes* and virtually all other members of the *Streptococcus* genus are known pathogens for a host of mammalian species, including our own. And in humans, a shocking number of illnesses are associated with this particular bacterium. *S. pyogenes* is one of the top ten causes of deadly infectious diseases for our species, and it's responsible for over half a million deaths annually. Among the diseases that can be chalked up to *S. pyogenes* are toxic shock syndrome, scarlet fever, strep throat, and a particularly scary one called necrotizing fasciitis, which has earned *S. pyogenes* the unpleasant epithet of "flesh-eating bacteria."

Work on *S. pyogenes,* therefore, has significant medical value, which makes it all the more enticing for researchers. In fact, it was Emmanuelle's desire to understand the pathogenicity of *S. pyogenes* that had led her to study CRISPR in the first place. The CRISPR system, she hoped, might give us new ways of tackling *Streptococcal* infections, saving countless lives.

Luckily for researchers, these virulent bacteria can be studied in ways that minimize their danger. When Emmanuelle approached me about collaborating, it was clear that my lab would focus on *S. pyogenes* exclusively in vitro (Latin for "within glass") as opposed to in vivo (Latin for "within the living"). We'd be doing test-tube experiments on purified proteins and RNA or DNA molecules instead of experiments with live cells and phages. We wouldn't need to grow cultures of *S. pyogenes* in sheep-blood-infused petri dishes or work in sealed-off laboratories to ensure the containment of this deadly pathogen. We'd also be able to use *E. coli,* our hardy laboratory workhorse, to mass-produce isolated genes and proteins from *S. pyogenes* safely, without risk of infection to the humans who handled them.

On my flight back to California after the Puerto Rico meeting, I thought about the proposed collaboration and wondered who in my lab I could ask to lead the project. By 2011, CRISPR research in my lab had grown considerably from its early days, and I now had several postdoctoral scientists, graduate students, and research specialists working on various aspects of CRISPR biology and tool development. But just about everyone was busy with his or her own project, and I was reluctant to foist new work on anybody.

Then it dawned on me that I had the perfect candidate: an extremely talented and hard-working postdoctoral scientist from the Czech Republic who was nearing the end of his time in my lab. He was already interviewing for faculty jobs but had recently mentioned that he was looking for something new to work on during his last year at Berkeley.

Martin Jinek (pronounced "*yeeh*-neck") was in many ways the opposite of Blake. Whereas Blake was outgoing and gregarious, Martin was reserved and introspective. When faced with an experimental obstacle or unfamiliar technique, Blake would immediately find someone who could help him; Martin would hit the books and figure it out for himself. If, that is, Martin didn't already have the answer in the first place. His knowledge of biology and biochemistry was encyclopedic, as evidenced not only by his lengthy and prestigious publication record but also by the diverse range of fields in which he'd published. Most important, he was familiar with the CRISPR field. After joining my lab with the intention of studying RNA interference in humans, he had also worked closely with Blake and Rachel to help complete a number of CRISPR-related projects.

Martin reacted enthusiastically when I pitched him the idea of collaborating with Emmanuelle. He suggested we also include Michael (Michi) Hauer, a master's student from Germany who was due to arrive in my lab in the summer. I agreed; the more hands on deck, the better. The more I'd learned about Emmanuelle's mystery protein, the more convinced I was that there really was something special there — something that might unlock the deepest secrets about CRISPR.

The Csn1 enzyme had gone by a variety of names over the years before one — Cas9 — finally stuck, in the summer of 2011. But as hard as it was to keep track of its changing aliases as I dug into the existing research about Cas9, I knew the protein's importance was beyond doubt. Rodolphe and Philippe's 2007 study had shown that inactivating the gene that coded for the Cas9 protein crippled *S. thermophilus*'s ability to protect itself against viral attack. Furthermore, when Josiane and Sylvain had discovered that phage genomes were being sliced apart during a CRISPR immune response, they'd also shown that inactivating the gene encoding Cas9 prevented CRISPR from destroying the viral DNA. Likewise, in Emmanuelle's experiments in *S. pyogenes,* mutations in the gene encoding Cas9 had caused defects in the production of CRISPR RNA molecules and also impaired the overall immune system. And finally, a study in the fall of 2011 from Virginijus Siksnys's lab, produced with Rodolphe and Philippe at Danisco, suggested that *cas9* was the only known protein-producing *cas* gene in the *S. thermophilus* CRISPR system that was absolutely essential for an antiviral response.

The more I read, the clearer it was that the Cas9 protein was likely to be a key player in the DNA destruction phase of the immune response in Type II CRISPR systems. At least, it seemed to be essential in bacteria within the *Streptococcus* genus, but it stood to reason that any critical component of one group of Type II systems would be just as vital in all the others. Exactly what role Cas9 was playing, however, remained to be determined.

Together with Martin, I had a Skype meeting with Emmanuelle to begin strategizing about our Cas9 experiments. Setting up the call had been a challenge, underscoring the logistical difficulties of our collaboration. Emmanuelle was at Umeå University in northern Sweden, ten hours ahead of Pacific standard time, and the PhD student leading the CRISPR project from her lab — Krzysztof Chylinski — was working out of the University of Vienna, where Emmanuelle's lab had previously been located. All told, we would be quite the international group: a French professor in Sweden, a Polish student in Austria, a German student, a Czech postdoc, and an American professor in Berkeley.

Once we finally found a time that worked for all of us, we began sketching the project in broad strokes. The initial goal, from my lab's perspective, was rather straightforward: we had to figure out how to isolate and purify the Cas9 protein, something that Emmanuelle's lab had been unable to do. With the Cas9 protein in hand, we could begin to conduct biochemical experiments aimed at determining whether Cas9 interacted with CRISPR RNA, as we suspected it did, and how it might function during an antiviral immune response.

Krzysztof, Emmanuelle's graduate student, shipped us an artificial chromosome containing the *cas9* gene from *S. pyogenes,* and Michi got to work on the protein purification under Martin's careful watch. First, Michi introduced the synthetic DNA into different strains of *E. coli,* and then, by varying the growth conditions and types of nutrient-rich broths, he systematically tested dozens of different parameters to find a single protocol that achieved the highest level of Cas9 protein production, much as a gardener might screen different soil and fertilizer combinations to identify the optimal growth conditions for a new flower. Next, using chromatography, a technique borrowed from chemistry, Michi tested different ways to crack open the cells and separate Cas9 from all the other cellular proteins. Finally, Michi tested the stability of the purified Cas9 protein. Some proteins are more finicky than others and "go bad" after just a single use, usually by aggregating and precipitating, much like microscopic snowflakes, causing a formerly clear tube of protein solution to turn milky white; others can be frozen and thawed repeatedly and exhibit excellent durability. We were in luck — Cas9 fell into the latter category.

Finally, it was time for us to run the first biochemical experiment. As Michi and Martin had been purifying and isolating the Cas9 protein, we'd envisioned that any DNA-cutting behavior the protein possessed would depend on the presence of CRISPR RNA. In the Type I CRISPR systems we had been studying in the lab, the CRISPR RNA assembled with a multitude of Cas proteins to form a DNA-binding-and-cutting machine. We imagined that Cas9 might work with CRISPR RNA in a similar fashion. Consistent with this idea, computational analysis of Cas9's amino acid sequence had revealed the likely presence of not one but two separate nucleic acid–cutting modules, or nucleases, within the enzyme. Maybe one or both of these modules were capable of cutting phage DNA.

With his time in the lab running out — he was expected back in Germany to complete his thesis and already had a flight booked — Michi, together with Martin, resolved to test whether the purified Cas9 enzyme was able to cut DNA. After synthesizing the functional CRISPR RNA molecule that Emmanuelle's work on *S. pyogenes* had defined, they mixed together the CRISPR RNA, the Cas9 protein, and a sample of DNA. Importantly, the DNA used in the experiment contained a sequence of letters that matched the sequence on one end of the CRISPR RNA.

As is so often the case in the scientific process, this experiment ended in disappointment. There was no observable change whatsoever to the DNA; it was precisely the same before and after being exposed to Cas9 and the matching CRISPR RNA. Either Michi hadn't yet set the experiment up properly or Cas9 didn't have the ability to cut DNA after all. Michi presented his results to the lab and somewhat dejectedly headed back home, thinking his summer of hard work spent isolating, purifying, and studying Cas9 had been for naught.

As our collaboration with Krzysztof and Emmanuelle got under way, Martin had been working closely with Michi and mentoring him in the lab, but he had also been preoccupied with his ongoing hunt for a faculty job. His interviews had taken him around the world, including to Switzerland, where he would ultimately accept a job as assistant professor at the University of Zurich. Fortunately for us, however, Martin's travel schedule had eased up considerably by the time Michi departed, so he was able to pick up where Michi had left off, turning his attention full-time to Cas9 — and to resolving the question of what exactly its function might be.

Michi and Martin's work seemed to show that Cas9 wasn't capable of cutting DNA, but could something have gone wrong with the experiment? There were many possibilities, from the mundane (for example, the protein might have degraded in the test tube) to the interesting (for instance, a necessary component of the reaction might have been missing). To explore this latter possibility, Martin and Krzysztof began testing different ways of setting up the DNA-cutting experiment. In one of the many serendipitous twists to the story, they soon discovered they had grown up just across the border from each other — Krzysztof in Poland and Martin in what was then Czechoslovakia — and both spoke Polish, which greatly helped the increasingly frequent Skype sessions they held to brainstorm about their experiments.

Eventually, Krzysztof and Martin performed experiments in which they included not only the CRISPR RNA but also the second type of RNA, called tracrRNA, that Emmanuelle's lab had found to be required for production of CRISPR RNAs in *S. pyogenes*. The result was simple but, to us, electrifying: DNA bearing a perfect match to twenty letters in the CRISPR RNA molecule was cleanly cut apart. Experiments conducted in parallel, called controls, showed that a match between the CRISPR RNA and the DNA sequence was essential, as was the presence of the Cas9 protein and the tracrRNA.

In essence, these results simulated what happens in a cell during a CRISPR immune response but with only the minimum of components; no cellular molecules besides Cas9 and the two RNA molecules, which looked similar to the way they'd look inside a *Streptococcus pyogenes* cell, along with a DNA molecule mimicking the genome of a phage. Of critical importance was the fact that twenty of those DNA letters matched those of the CRISPR RNA, meaning that the CRISPR RNA and one of the two DNA strands should be able to form their own double helix through complementary base pairing. Such an RNA-DNA double helix could be the key to the specificity of Cas9's DNA-cutting activity.

Monitoring the DNA-cutting reaction in a test tube required a sensitive detection method, since there was no way to directly visualize the DNA being cut. At fifty letters long, the DNA double helix would be just seventeen nanometers, or seventeen-billionths of a meter, long, roughly one-thousandth the width of a human hair. Not even the most powerful microscope could show us that, so Martin and Krzysztof employed two of the favorite tools of a nucleic acid biochemist: radioactive phosphorus and gel electrophoresis. Radioactive phosphorus atoms were chemically attached to the ends of the DNA molecules in these experiments, ensuring that the DNA would light up when exposed to x-ray film, and then a strong electrical voltage was used to force all the DNA through a large slab of Jell-O-like material that acted like a molecular sieve, separating the molecules based on their size. Exposing the x-ray film to the gel revealed multiple regions, or bands, of signal once the DNA was cut by

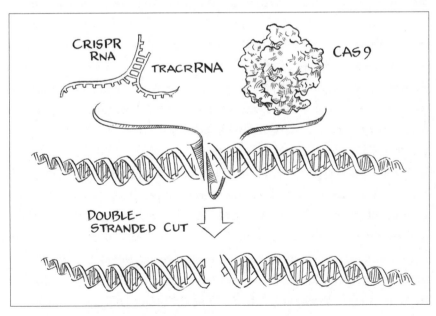

Cas9 cuts DNA using two RNA molecules

Cas9 — one band for the full-size DNA and another for the DNA that had been sliced into two fragments.

Martin further showed that both DNA strands were being cut by the Cas9 protein at the same position relative to the position of the CRISPR RNA. Importantly, the CRISPR RNA and tracrRNA molecules remained the same throughout the experiment and so could be used repeatedly by the Cas9 protein to determine the DNA sequence to be cut.

Looking at these results, we realized that we had defined the essential parts of the DNA-cutting machine, the mechanism that allowed *S. pyogenes* and *S. thermophilus* — and any other bacteria with a similar type of CRISPR system — to not only target specific phage DNA sequences, but also to destroy them. The crucial components for DNA cutting were the Cas9 enzyme, the CRISPR RNA, and the tracrRNA.

I was elated by these results but also consumed by the multitude of follow-up questions we urgently wanted answered. To understand exactly how the Cas9 enzyme was capable of RNA-guided DNA cutting, we needed to pinpoint the part of the protein that controlled its cutting function. To prove that DNA cutting was specific and required a match between the CRISPR RNA and DNA sequence, we needed to vary the letter-by-letter sequence of the DNA and show that cutting was prevented when the RNA-DNA match was imperfect. And to demonstrate how the CRISPR RNA and tracrRNA molecules worked, we needed to systematically pare down both RNA molecules and figure out what parts of the RNA molecules were truly necessary.

Martin and Krzysztof labored tirelessly to answer these questions, and, quickly, an incredible picture began to emerge. They found that Cas9 could latch onto a DNA double helix, pry open the two strands to form a new helix between the CRISPR RNA and one strand of DNA, and then use two nuclease modules to simultaneously slice through both strands of the DNA, creating a double-strand break. Depending on the sequence of its associated RNA molecule, Cas9 could target and cut virtually any matching DNA sequence. In effect, the CRISPR RNA molecule acted like a set of GPS coordinates, guiding Cas9 to a precise spot within the vast expanse of a long DNA molecule according to the matching letters in the CRISPR RNA and DNA. Here was a truly programmable nuclease, one that would be able to target any arbitrary DNA sequence using the same base-pairing rules — A goes with T, G goes with C, and so forth. For any twenty-letter sequence the guide RNA contained, Cas9 would find its matching counterpart in DNA and then cut.

In the warfare waged between bacteria and viruses, Cas9's function made perfect sense. Armed with a cache of RNA molecules derived from the CRISPR array, where snippets of phage DNA had been stored, Cas9 could readily be programmed to slice up corresponding sites within viral genomes. It was the perfect bacterial weapon: a virus-seeking missile that could strike quickly and with incredible precision.

With Martin's and Krzysztof's results in hand, we were ready to tackle the next question: If bacteria could program Cas9 to cut up specific viral DNA sequences, could we, the researchers, program Cas9 to cut up other DNA sequences — viral or not — as we suspected? Martin and I were keenly aware of the developments in the gene-editing field and of the promise — but also the serious limitations — of the ZFN- and TALEN-based programmable nucleases. We realized, with no small sense of awe, that we had come upon a system that could be transformed into a far more straightforward gene-editing technology than anything previously discovered or developed.

To turn this tiny molecular machine into a powerful gene-editing tool, we'd have to take one more step. So far, we had reduced a complex immune response into a simple set of moving parts that could be isolated, modified, and combined in different ways. What's more, through careful biochemical experiments, we had deduced the molecular rules governing the functions of these different parts. What we wanted to do next was confirm that we could engineer Cas9 and the RNA molecules to target and cut any DNA sequence of our choice. This demonstration would highlight the full power of CRISPR.

That single step — programming the CRISPR-Cas9 machine ourselves — actually consisted of two smaller steps: developing an idea, then performing an experiment.

First came the idea. Meticulous as ever, Martin had systematically modified both RNA molecules — the CRISPR RNA molecule that did the targeting, and the tracrRNA molecule that held it and Cas9 together — to determine how every letter in each RNA affected function. Using this knowledge, Martin and I brainstormed a way to convert the two RNA molecules into one. If we fused the tail of one to the head of the other, the resulting chimeric RNA, if functional, would simplify our programmable DNA-cutting machine; instead of having to combine Cas9 with both RNA molecules, the guide (CRISPR RNA) and the helper (tracrRNA), we'd be able to pair the enzyme with a single RNA molecule — a single-guide RNA — that did both jobs. In making CRISPR a tool for gene editing, reducing the complexity of the system would go a long way toward increasing its usability.

Based on this idea, we designed an experiment. We needed to test this single, fused RNA molecule and determine whether it could still guide Cas9 to cut a matching DNA sequence. In addition, our experiment would indicate whether Cas9 could indeed be programmed to cut any DNA sequence we wanted it to, as we suspected, and not just phage DNA sequences that CRISPR had naturally selected over the course of bacterial evolution.

Driven more by convenience than by preference — at this point, we knew we had a major breakthrough and didn't want to delay our experiment by picking a gene that wasn't already in the lab's freezer — we decided to target a jellyfish gene called green fluorescent protein, or GFP. (Widely used in labs worldwide to visualize cells and their component proteins, GFP became such an important biotechnology tool that it

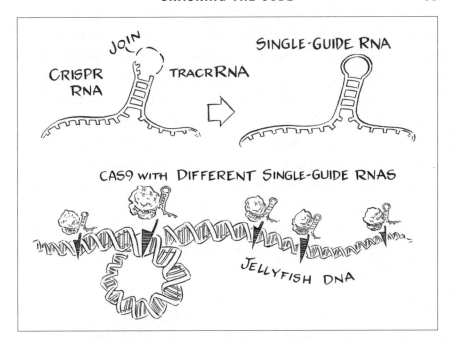

Programmable DNA cutting by CRISPR-Cas9

earned Martin Chalfie, Osamu Shimomura, and Roger Tsien the 2008 Nobel Prize in Chemistry.) Martin Jinek chose five different twenty-letter sequences within the gene to target and then engineered five chimeric RNA molecules to precisely match them. Once the new single-guide RNA molecules had been prepared, we incubated them with Cas9 and the jellyfish DNA in the same DNA-cleavage assay that had by now become routine and then waited for the results.

As Martin walked me through the data from the GFP experiment while we stood at one of the computers in the lab, I saw a gel scan that looked beautiful. All of the GFP DNA had been sliced apart at the intended sites. Each single-guide RNA molecule had worked as it was meant to, selecting the exact spot in the jellyfish DNA that we wanted cut and pairing with Cas9 to break it apart at that precise location.

We'd done it. In a short time, we had constructed and validated a new technology that, based on the body of research conducted with ZFN and TALEN proteins, would be capable of editing the genome — any genome, not just one belonging to a bacterial virus. Out of this fifth bacterial weapons system, we had built the means to rewrite the code of life.

That night, as I stood at my kitchen stove cooking dinner, visions of this tiny machine danced in my head: Cas9 and its guide RNA whizzing around a bacterial cell, hunting for matching DNA sequences. Suddenly, I found myself laughing aloud. How incredible that bacteria had found a way to program a warrior protein to seek and destroy viral DNA! And how miraculous, how fortunate, that we could repurpose this fundamental property for an entirely different use. It was a precious time of pure joy, the joy of discovery — a feeling just like I'd felt in Dr. Hemmes's lab all those years before.

In June 2012, Emmanuelle and Krzysztof came to Berkeley for a conference, giving Martin and me a chance to finally see them in person again. Incredibly, given the scientific distance all of us had traveled together, our communication up until that point had been almost exclusively virtual. After innumerable phone calls, Skype discussions, and e-mail exchanges, we were all sitting in my Berkeley office, marveling at the results of our brief but intense collaboration.

Emmanuelle and Krzysztof were in town for the fifth annual CRISPR conference, a meeting that brought together members of the twenty or thirty laboratories that were studying the bacterial defense system in 2012. Most of these researchers were in the fields of food science and microbiology. CRISPR still hadn't received much notice from the larger scientific community; only about two hundred scientific articles mentioning the system had been published in the prior decade. That, we knew, was about to change.

The timing of the meeting couldn't have been better — or worse. On the one hand, we would be able to compare our work with the work of our colleagues. On the other hand, the previous few weeks had been absolutely, punishingly frenetic, and all of us were ready for a break.

Following our GFP experiment, we had decided to wrap up the project as fast as possible and put together a research article describing it. Even before Martin and Krzysztof had finished all their experiments, and before our foreign collaborators had started packing for their trip to Berkeley, Emmanuelle and I began writing.

Our article focused primarily on how our work explained the functioning of CRISPR for antiviral defense in *S. pyogenes,* but we also wanted to point out the profound implications of our results. We included a statement in the abstract of the paper pointing out the utility of a programmable DNA-cutting enzyme for genome editing. In addition, we concluded the article with a brief but significant nod to uses of CRISPR outside of bacteria, including in other cell types. After mentioning ZFNs and TALENs, we wrote in our concluding sentence: "We propose an alternative methodology based on RNA-programmed Cas9 that could offer considerable potential for gene-targeting and genome-editing applications."

On June 8, 2012, a sunny Friday afternoon, I clicked Confirm on my computer, formally submitting our paper for consideration to the journal *Science.* It would be published just twenty days later, on June 28, and nothing after that would ever be the same — not for me, not for my collaborators, and not for the field of biology. In that moment, however, my elation at this milestone was muted. I was more exhausted than I'd ever been in my life.

I felt like I had been sitting at my desk for weeks, so I stood up and — in a bit of a daze — wandered out of Stanley Hall and onto the green expanse of the Berkeley campus. The lawn around the circular reflecting pool in front of our building was conspicuously empty. The spring semester had ended almost a month earlier; the campus, usually bustling, was eerily quiet.

It was, I now realize, the calm before the storm.

4

COMMAND AND CONTROL

ROUGHLY A YEAR AFTER our CRISPR article was published in *Science*, I found myself in Cambridge, Massachusetts, on the first of what would turn into monthly trips across the country to discuss our invention.

It was early June 2013, and I had traveled to the campus of Harvard University to meet with a rising scientist in Harvard's Department of Stem Cell and Regenerative Biology. Professor Kiran Musunuru's office was in the Sherman Fairchild Building, where I had attended biological chemistry lectures as a graduate student back in the 1980s. Thirty years later, the building looked much the same from the outside. Inside, however, it was completely renovated. Gone were the antiquated lecture halls and dated biochemistry labs; in their place was a white, airy, well-lit facility housing state-of-the-art equipment. On the day I visited, this modern space was alive with dozens of researchers working together to plumb the deepest mysteries of cell and tissue growth.

In some ways, Fairchild's transformation—especially its shift of focus from basic biochemistry to applied biology—mirrored a conceptual metamorphosis in my own mind and work. The past year had been a whirlwind. Researchers all around the world, far too many scientists to fit into any one lab, had rapidly exploited our revelations about the biochemical properties of CRISPR-Cas9. They were already using this new knowledge to engineer the DNA of countless organisms as well as the genetic material in human cells. Academics and physicians alike were hailing CRISPR as the holy grail of gene manipulation: a quick, easy, and accurate way to fix defects in genetic code. In what felt like the blink of an eye, I had been transported from the field of bacteria and CRISPR-Cas biology to the world of human biology and medicine. For a highly specialized academic like me, it was a quantum leap, like falling asleep in Berkeley and waking up on Mars.

My meeting with Kiran perfectly exemplified the excitement around this new technology. I had come to Harvard to discuss the applications of CRISPR as a therapeutic tool, but Kiran was already one step ahead of me. Before we sat down in his office, he beckoned me to follow him into the lab. While we walked, he gave me an animated overview of the myriad ways in which his research team was using CRISPR to develop treatments for genetic diseases.

One of the targets of his team's research, Kiran explained, was sickle cell disease — the condition in which a single DNA mutation interferes with red blood cells' ability to ferry oxygen through the body. Members of his lab had been using CRISPR to target and cut the mutated beta-globin gene, catalyzing reversion of the faulty letter A at position 17 back to the correct letter, T. If they could refine this technique in actual human cells in the laboratory, there was every reason to believe that the same feat could one day be performed in patients, forming the basis of a treatment that would eliminate the genetic disease at its source.

I followed Kiran to a computer workstation in which long strings of DNA sequences were displayed from left to right, one on top of the other. He pointed to the two sequences at the top and explained that these were the letters of beta-globin DNA from blood cells taken from two patients, one healthy, the other suffering from sickle cell disease. Sure enough, the normal individual possessed a T in the seventeenth position whereas the diseased individual had an A in that spot.

Then Kiran guided me to the bottom-most panel. The DNA sequence there was also from cells taken from the sickle cell patient, but after they had been infused with three CRISPR-related elements. The first was a set of genetic instructions to produce the Cas9 protein from *Streptococcus pyogenes*. The second element was a CRISPR-derived guide RNA molecule designed to home in on the beta-globin gene precisely at the mutated site. The third element was a synthetic DNA replacement — a piece of the healthy beta-globin sequence that cells would use to patch up and repair the gene after Cas9 had gone in and sliced it apart. Kiran was using the CRISPR-Cas9 system, or CRISPR for short, to target the right part of the genome and cut it, but also to mark it for repair; the cell would then replace the faulty sequence with the correct replacement sequence.

Looking at the bottom of the computer screen, I was delighted to see that this was exactly what had transpired: the DNA sequence from the sickle cell patient now looked indistinguishable from the sequence taken from the healthy patient. Using CRISPR, Kiran's team had perfectly swapped out the disease-causing letter A for the normal letter T without disturbing the genome in any other way. In one simple experiment using a patient's blood cells, they had shown that the CRISPR-Cas9 system was capable of curing a crippling disease affecting millions of people worldwide.

That evening, I went for a run along the Charles River. I'd taken this route many times as a graduate student, and now, with the familiar waters of the Charles rushing beside me, I felt like I was back in school once more. As I jogged, my mind flashed back to conversations I'd had about DNA repair during my graduate years, discussions that centered on research published by my adviser, Jack Szostak, and his graduate student Terry Orr-Weaver. At the time, many scientists were puzzled by their model describing how cells repaired damage to the DNA double helix, and they were even more flummoxed by the idea, advanced by Memorial Sloan Kettering's Maria Jasin and others, that researchers could harness this mode of repair to alter specific DNA sequences. But this strategy had worked well for previous technologies like the ZFN and TALEN enzymes, and we were now witnessing the same strategy working well with CRISPR. What's more, the CRISPR-based gene-editing method was much easier to employ. Would CRISPR supplant these older technologies the same way that compact discs had replaced cassette tapes (and cassette tapes had replaced vinyl)? My musings took me all the way from Harvard Square to the Longfellow Bridge and back, the passing cityscape a mere blur as thoughts of CRISPR raced through my mind.

What really excited me was the thought that the approach Kiran's lab had taken could be applied to a host of other genetic diseases. If scientists could safely and efficiently deliver CRISPR into the human body so that gene editing worked as well in patients as it did in lab-cultured cells, then the possibilities for transforming medicine would be boundless. To make good on this promise, though, would require resources and human-power far greater than any academic laboratory could furnish alone. It was for this reason that some colleagues and I were considering founding a company to develop CRISPR-based therapies — this was the purpose of my visit to Cambridge, in fact. Our dream was to leverage CRISPR to treat genetic diseases in a way that had scarcely been feasible before.

In a marathon of meetings in the summer and fall of 2013, the team of this hypothetical company evolved to include me and four other scientists: George Church, Keith Joung, David Liu, and Feng Zhang. In November 2013, we founded Editas Medicine with $43 million in financing from three venture capital firms. Just a half a year later, Emmanuelle co-founded another company, CRISPR Therapeutics, with an initial $25 million bankroll, and in November 2014, a third company, Intellia Therapeutics, joined the scene with $15 million in Series A funding. By the end of 2015, these three companies would raise well over half a billion dollars more for research and development of therapies to target numerous disorders, from cystic fibrosis and sickle cell disease to Duchenne muscular dystrophy and a congenital form of blindness, all using the CRISPR technology that Emmanuelle and I had first developed and described.

As exciting as the medical possibilities were, it would still take years for CRISPR to make its way to human clinical trials. In the meantime, the technology quickly disseminated through the global scientific community as word spread that gene editing inside living cells could now be performed easily within days. Many experts predicted that CRISPR would be a research biologist's dream come true, enabling experiments that one could have only fantasized about doing before. I imagined that it would democratize a technology that had once been the privilege of the few. In the days before CRISPR, gene editing required sophisticated protocols, formidable scientific expertise, and substantial financial resources, and it could be performed on only a few model organisms. By the time of my first visit to Harvard, though, even laboratories with no prior gene-editing experience were using the technology.

The days of mentioning CRISPR at a seminar or scientific conference and receiving mostly blank stares were long gone. Now, CRISPR seemed to be on everyone's lips and the topic of every conversation. And yet it was still only the tip of the iceberg.

As I sat on the plane flying back to San Francisco after that first trip to Cambridge, I could already see a new era of genetic command and control on the horizon — an era in which CRISPR would transform biologists' shared toolkit by endowing them with the power to rewrite the genome virtually any way they desired. Instead of remaining an unwieldy, uninterpretable document, the genome would become as malleable as a piece of literary prose at the mercy of an editor's red pen. As I contemplated these vast possibilities, I could hardly believe how quickly things had evolved since Martin's and Krzysztof's initial successes programming CRISPR to slice up DNA in a test tube. Now, the scientific community stood in a rapidly expanding pool of light — one that was growing to reveal incredible new insights into how CRISPR worked and how it would someday be used to improve human health.

In their experiments published in our 2012 *Science* article, Martin and Krzysztof had demonstrated something groundbreaking: that a

CRISPR-associated protein called Cas9, isolated from flesh-eating bacteria, worked with two molecules of RNA to target matching twenty-letter DNA sequences and cut them apart. The RNA acted like a guide, dictating the GPS coordinates of the attack, and Cas9 acted like the weapon to eliminate the target. In bacteria infected by a virus, this CRISPR machine was mobilized to slice up and destroy specific DNA molecules from the virus as part of an adaptive immune response.

Virginijus Siksnys and colleagues published a similar paper to ours in the fall of 2012 describing the function of the Cas9 protein found in yogurt-producing bacteria — a member of the same *Streptococci* genus. Like us, they found that Cas9 cut apart DNA sequences that matched the letters of the CRISPR RNA. But they failed to uncover the crucial role of the second RNA (called tracrRNA), which we had demonstrated was an essential component of the DNA-targeting and DNA-cutting reaction.

In our paper, we had described the molecular requirements of this defense system in exhaustive detail and shown how easy and simple it was to design new versions of CRISPR to cut up any DNA one chose. Yet we had gone a step further and reengineered the RNA guide, made up of two separate RNA molecules in bacteria (CRISPR RNA and tracrRNA), into a single-guide RNA molecule that still enabled Cas9 to find and cut a particular DNA sequence. We had also proposed that this defense system could be repurposed for a different function inside cells, not to destroy viral DNA, but to precisely edit the cell's DNA. If we changed the twenty-letter RNA code to match the sequence of a specific human gene and then transplanted Cas9 and the new guide RNA into human cells, CRISPR would make a surgical cut in the targeted gene, marking that site for repair. By slicing the DNA apart, CRISPR would be acting like a red alert triggering the cell to fix the damage, but in a way that we could control.

Using CRISPR in human cells, as we had proposed, would confirm the power of this new mode of gene editing. And there were good reasons to expect success. Our own research had revealed that the Cas9 protein and its guide RNA were very picky about their partners and stuck together tightly, indicating they should have no problem finding each other inside a human cell. As for sending them into the cell's nucleus, where the DNA is located, we could simply provide a chemical zip code that would let the cell do the work for us. Plenty of labs before us had succeeded in transplanting proteins and RNA molecules from bacteria to human cells, and there were many molecular tools at our disposal that we could use to help CRISPR work efficiently outside its natural environment.

We just had to show that it worked as expected.

Martin began by transferring the bacterial DNA encoding Cas9 and the CRISPR-derived RNA into two plasmids, little ringlets of DNA that act like artificial mini-chromosomes. The first plasmid contained genetic instructions for the guide RNA as well as separate instructions that directed human cells to produce gobs of it. The second plasmid contained the *cas9* gene, but it had been "humanized" so that it could be interpreted by protein-synthesizing factories inside human cells. Martin also fused the *cas9* gene to two genes, routinely used by biologists, that coded for other proteins: a tiny one, called a nuclear localization signal, which directed a protein to the cell's nucleus, and the green fluorescent protein that would cause any human cells successfully producing Cas9 to fluoresce green when exposed to ultraviolet light.

By combining all these molecular parts, Martin and I intended to convert human cells into CRISPR-producing factories that unwittingly churned out molecules programmed to target and cut up their own genome. Yet we knew that CRISPR wouldn't destroy human cells by cutting their DNA the way it destroyed viruses in bacteria by cutting viral DNA. Humans (all eukaryotic organisms, for that matter) constantly suffer DNA damage — it occurs when we are exposed to carcinogens, UV light, or x-rays, for example — and to mend damaged DNA, cells have evolved intricate DNA-repair systems that fix double-strand breaks. Thus, in the most basic scenario, if CRISPR succeeded in cutting a gene, the cell would respond by simply gluing the DNA back together, much like welding two pieces of metal pipe together. Scientists refer to this process as nonhomologous end joining, since, unlike homologous recombination, the mending does not involve a matching repair template. (*Homologous* derives from the Greek *homologos,* meaning "agreeing.")

A key property of this repair process is its inherent sloppiness. Just as a welder needs to be sure that the two pipes have clean edges before he or she joins them, the cell needs to ensure that the broken pieces of DNA have clean ends before putting them back together. Generating clean ends sometimes involves deleting or inserting a few letters of DNA, which results in permanent genetic changes after the repair process is completed. This meant that a gene would most likely be altered after being targeted by CRISPR, sliced apart, and repaired by the cell. This messy, error-prone repair would give Martin and me a simple way to detect successful gene editing. By targeting a specific gene and analyzing its DNA sequence letter by letter before and after CRISPR treatment, we would find any signs of sloppy repair, thereby proving that CRISPR had located and cut its target.

Martin and I decided to program CRISPR to target a human gene called clathrin light chain A, or *CLTA,* which plays a role in endocytosis, a process that cells use to internalize nutrients and hormones. We weren't studying endocytosis, but the *CLTA* gene had already been edited with the older ZFN technology by Professor David Drubin's laboratory, also at Berkeley. Thus, we knew that editing this gene was possible and that testing CRISPR side by side with ZFN would allow us to compare and contrast them. Of course, building the ZFN tool to edit *CLTA* had required several months and a crucial collaboration with a company that provided David's team with the ZFNs free of charge (the going rate at the time — twenty-five thousand dollars per ZFN — was prohibitively expensive). In stark contrast, it took Martin just minutes sitting at his computer to design the analogous version of CRISPR, and it could be generated for a few tens of dollars. This was, after all, one of CRISPR's greatest attributes — that it was fantastically easy to target specific genes.

All you had to do was select the desired twenty-letter DNA sequence to edit and then convert that sequence into a matching twenty-letter code of RNA. Once inside the cell, the RNA would couple with its DNA match using base pairing, and Cas9 would slice apart the DNA.

The actual proving ground of our first gene-editing test would be a line of human embryonic kidney cells called HEK 293. First generated in 1973 from kidney cells obtained from an aborted fetus, HEK 293 cells had gained popularity among cell biologists because of how easily they could be cultured in the laboratory and how readily they accepted foreign DNA. When we mixed the two plasmids — one with genetic instructions to make Cas9 and the other with genetic instructions to make guide RNA — in a soapy solution of molecules called lipids, the mini-chromosomes (plasmids) would be spontaneously engulfed by microscopic greaseballs, just like the fat globules that float on the surface of a chicken soup. After we added this mixture to HEK 293 cells, the greaseballs would merge with the cell membrane and dump the DNA contents into the cell's interior. Once inside, the DNA would be copied, transcribed, and translated, producing the Cas9 protein and *CLTA*-specific guide RNA. The DNA-cutting machine would then have to make it inside the nucleus, where our target DNA sequence was housed. It would need to locate and cut the correct twenty-letter DNA sequence. And the cell would have to repair the broken DNA in such a way that we could detect it.

Martin's experiments immediately demonstrated that the mini-chromosomes were indeed allowing human kidney cells to generate the CRISPR components. When Martin inspected the cells under a microscope, he observed a high percentage of cells glowing green, which could result only from production of Cas9 together with the green fluorescent protein fusion. After collecting a portion of the cells and grinding them up to analyze the different RNA molecules inside, Martin also found that the kidney cells were churning out copious amounts of the guide RNA.

The transplantation of CRISPR from bacterial cells to human cells

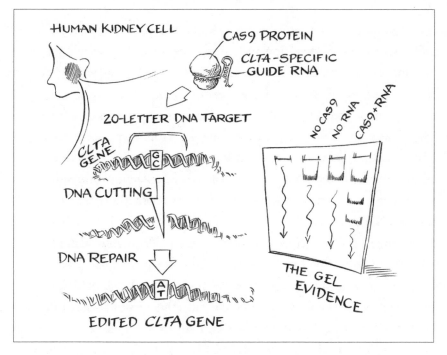

Editing DNA in human cells with CRISPR

had worked as we had expected, leaving one last obvious question: Was CRISPR editing human DNA?

Martin and a young student who had recently joined the project, Alexandra East-Seletsky, ground up some more cells, extracted the DNA, and analyzed the *CLTA* gene. The answer was unmistakable: the gene had been edited at exactly the site matching the CRISPR RNA sequence. To the untrained eye, the results didn't look like much — a bunch of dark bands on a thin slab of gel-like material — but the implications were enormous.

In just a few simple and routine steps, Martin and I had selected an arbitrary DNA sequence within the 3.2-billion-letter human genome, designed a version of CRISPR to edit it, and watched as the tiny molecular machinery followed through with its new programming — all inside living human cells. With that success, we had validated our new technology that offered scientists the remarkable ability to rewrite the code of life with surgical precision and astonishing simplicity. In what felt like no time at all, CRISPR had already caught up to almost twenty years of research and development in other gene-editing technologies.

In a virtual repeat of our rush half a year earlier to publish our research with Krzysztof and Emmanuelle, we wrote a manuscript describing our latest results. Whereas our first 2012 manuscript had contained a clear directive to apply CRISPR in cells as a novel gene-editing platform, our second was a clear demonstration and confirmation of the power of this newfound system.

As 2012 came to a close, I viewed it with considerable irony that *Science* magazine, which had published our previous CRISPR paper just six months prior, named genome editing one of the runner-up breakthroughs of the year (first prize went to the Higgs boson) but highlighted an older technology — TALEN — that had been discovered just before our work with CRISPR. I wondered what else CRISPR might have in store for the scientific community.

Much to my delight, the first few weeks of 2013 were marked by the publication of a whopping five articles on CRISPR besides our own, all describing similar kinds of experiments in which the system had been used to edit genes in cells, just as we had proposed in 2012. Both MIT professor Feng Zhang and Harvard professor George Church had contacted me to alert me to their forthcoming publications. The Zhang and Church articles in the journal *Science* appeared online in early January, followed later that month by my article with Martin and three other articles from the labs of Professor Jin-Soo Kim of Seoul National University, Rockefeller University professor Luciano Marraffini, and Harvard Medical School professor Keith Joung.

It was a heady time. I was elated that the work published with Emmanuelle the preceding summer had inspired others to pursue a line of experimentation similar to our own. Only later would the contents and publication dates of these papers be dissected to support arguments in a dispute over CRISPR patent rights, a disheartening twist to what had begun as collegial interactions and genuine shared excitement about the implications of the research.

When I compared the six articles, I realized that over a dozen different genes had been edited. Even more exciting than the diversity in edited DNA sequences, though, was the diversity in edited cell types. In addition to editing genes in embryonic kidney cells, CRISPR had been programmed to slice up DNA in human leukemia cells, human stem cells, mouse neuroblastoma cells, bacterial cells, and even one-cell embryos from zebrafish, a popular model organism for genetics studies. CRISPR wasn't just showing some signs of success; it was exhibiting incredible versatility. As long as the Cas9 protein was present and the guide RNA had a twenty-letter code that matched a twenty-letter DNA code, it seemed that virtually any gene in any cell could be targeted, cut, and edited.

The excitement surrounding CRISPR intensified in May, when Rudolf Jaenisch's lab at MIT reported the generation of gene-edited mice using CRISPR. Just six years earlier, the Nobel Prize in Physiology or Medicine had been awarded to several scientists for developing methods to introduce genetic changes in mice, the most widely used animal model for the study of mammalian genetics. For over two decades, this effective but laborious method had been the best way — really, the only way — of replicating human disease–causing or cancer–causing mutations in mice. Back in 1974, Jaenisch himself had been the first to create a transgenic mouse containing foreign genetic material, and fifteen years later, he made headlines again by being one of the first to adopt the Nobel Prize–winning technique. But now, Jaenisch's success with CRISPR spotlighted a new technology that not only supplanted the old approach but suggested a way to seamlessly edit the genome of other animals too.

The previous gene-targeting method required embryonic stem cells, extensive backcrossing or interbreeding, and many generations of mice;

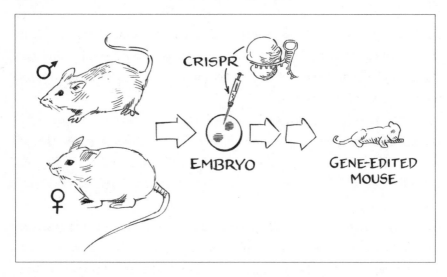

Creating gene-edited mice with CRISPR

it wasn't uncommon for an entire PhD thesis to culminate in the production and characterization of just a single genetically modified mouse
strain. Jaenisch's team employed CRISPR to achieve the same feat in
just one month, using a simple, streamlined protocol: microinjection
of CRISPR components directly into one-cell embryos, followed by implantation of the gene-edited embryos into a female's womb. Moreover,
they showed that CRISPR could be programmed with not just one RNA
guide, but multiple different guides, directing Cas9 to cut up and edit
several DNA sequences in mouse embryos simultaneously. This kind of
one-step, multiplex gene editing had never before been performed in
mice.

What was arguably most exciting about Jaenisch's study — at least for
geneticists working with animals other than mice — was that it revealed
an almost effortless way to edit genes in nearly any organism. Whereas
the original technique employing embryonic stem cells was used only
in mice, it seemed that CRISPR could be injected into any species' germ
cells (eggs and sperm) or embryos, and the resulting genetic changes

would be faithfully copied into all the cells and forever transmitted to future offspring. I did not envision at that time that extending CRISPR's use to human embryos would spawn one of the biggest controversies surrounding CRISPR — one in which I would soon be swept up.

In the summer of 2013, as I was marveling at the pace of CRISPR's dissemination, I began keeping a list of all the different cell types and organisms whose genomes had been edited using the technology. The list was manageable at first and included the zebrafish and cultured bacterial, mouse, and human cells from January and February, followed by yeast, mice, fruit flies, and microscopic worms. At the end of that year, my list included rats, frogs, and silkworms. By the end of 2014, I had added rabbits, pigs, goats, sea squirts, and monkeys, after which, as I acknowledged to audiences at seminars where I shared the list, I honestly lost track. Witnessing the protein and RNA molecules naturally deployed as antiviral defenses in bacteria being used to snip apart and precisely edit DNA sequences across the animal kingdom was breathtaking.

And it wasn't just animals. Plant biologists, while a bit slower on the initial draw, were revealing the incredible potential of CRISPR to edit DNA in crops and other plant species. A bout of publications in the fall of 2013 reported the successful use of CRISPR for gene editing in staples such as rice, sorghum, and wheat, and a year later, the list of plants had expanded to include soybeans, tomatoes, oranges, and corn.

The list of CRISPRized plants and animals has continued to grow. As of 2016, scientists have edited DNA in everything from cabbage, cucumbers, potatoes, and mushrooms to dogs, ferrets, beetles, and butterflies. Even viruses, those biological entities straddling the boundary between animate and inanimate matter — as they lack the ability to replicate autonomously but still possess genetic material — have had their genomes rewritten using CRISPR, the very same bacterial system that evolved to destroy them.

Meanwhile, though adult *Homo sapiens* are among the last animals to be treated with CRISPR, human cells have been subjected to more CRISPR gene editing than those of any other organism. Scientists have applied CRISPR in lung cells to correct the genetic mutation that causes cystic fibrosis, in blood cells to correct the mutations that cause sickle cell disease and beta-thalassemia, and in muscle cells to correct the mutations that cause Duchenne muscular dystrophy. Scientists have used CRISPR to edit and repair mutations in stem cells, which can then be coaxed to transform into virtually any cell or tissue type in the body. And scientists have used CRISPR to edit thousands of genes in human cancer cells in an attempt to discover new drug targets and new therapies.

If there was anything more exciting than watching CRISPR used in just about every species imaginable, it was seeing the very limits of gene editing being stretched and expanded. Back in the 1980s, scientists had been content to edit individual genes at efficiencies that were just fractions of a percent. By the early 2000s, the efficiencies moved into the low-single-digit percentages, and it became possible to alter genes in a couple of new ways. But with CRISPR, gene editing was now so powerful and multifaceted that it was often referred to as genome *engineering*, a reflection of the supreme mastery that scientists held over genetic material inside living cells.

In the process of using CRISPR on a range of different organisms, scientists have developed and refined a multitude of tactics for editing DNA. In addition to simply slicing apart DNA and inserting new sequences into the target genome, they can now also deactivate genes, rearrange sequences of genetic code, and even correct single-letter mistakes, as Kiran Musunuru had demonstrated during my visit to his lab. These advances have, in turn, allowed scientists to perform new types of experiments in the plant and animal kingdoms, including in our own species. So before we go any further into these applications of gene editing, it's important to understand the many potential uses of this incredibly versatile tool.

• • •

In the spring of 2014, my son Andrew's sixth-grade science teacher asked me to visit the classroom and explain CRISPR to the students. I was honored to be invited but also quite nervous: How would I describe gene editing to a group of kids who had only a basic knowledge of DNA?

I decided to bring along a 3D-printed model of the Cas9 protein and its guide RNA attached to DNA. This model had become a centerpiece in my office, its electric-orange RNA and brilliant blue DNA entwined with the snow-white protein in a football-size unit held together by magnets. The underlying molecular details might be a bit much for the kids, I thought; I figured I'd just pass around the football so they could look at it up close.

I underestimated the students' curiosity. Almost as soon as I handed them the model, they figured out how to break the DNA where Cas9 cuts it and how to pull the DNA in and out of the CRISPR assemblage. So much for worrying about communicating a complicated concept!

As I explained to the class, CRISPR can be described as a pair of designer molecular scissors because of its core function: to home in on specific twenty-letter DNA sequences and cut apart both strands of the double helix. Yet the types of gene-editing outcomes that scientists can achieve with this technology are remarkably diverse. For this reason, it might be better to describe CRISPR not as scissors but as a Swiss army knife, a tool with a panoply of functionalities that all stem from the action of a single molecular machine.

The simplest use of CRISPR is also the one that's most widely employed: have it cut a specific gene and then allow the cell to repair the damage by reconnecting the strands. This sloppy, error-prone process leaves telltale clues — short insertions or deletions of DNA (known as indels) flanking the sequence cut apart by CRISPR. Even though scientists can't control the exact way that the DNA is repaired in this particular use of CRISPR, they've realized how useful this type of gene editing can be.

Genes are, after all, just carriers of information, like the blueprints of a house; the goal of gene editing is not merely to alter the blueprints, but to change the form of the structure that gets built. In many cases, this means altering the proteins that the genes encode and that cells produce during gene expression.

Gene expression is the process by which the simple letters of DNA are translated into functional proteins, according to the central dogma of molecular biology. First, a temporary copy of DNA, called messenger RNA (or mRNA), is made in the cell's nucleus. Like a strand of DNA, the mRNA is a chain of letters, and its sequence matches the sequence of the DNA it copied (the only major exception being that T gets replaced by U). The mRNA is exported out of the cell's nucleus and delivered to a protein-synthesizing factory called a ribosome, which translates the four-letter language of RNA (A, G, C, and U) into the twenty-letter language of proteins (the twenty amino acids). This translation proceeds according to the genetic code, a cipher in which every three-letter RNA combination, called a codon, instructs the ribosome to add one specific amino acid. (With sixty-four possible codons but only twenty amino acids, many codons code for the same amino acid, and three codons serve as stop signs to terminate protein synthesis.) The ribosome begins at one end of the mRNA and reads one consecutive codon after another, adding the corresponding amino acids to the growing protein chain until it reaches the other end of the mRNA; the process is much like building a car on an assembly line. A critical feature of this system is that the ribosome *must* remain in the correct three-letter reading frame; even the slightest hiccup can dramatically and calamitously affect the entire translation.

To better appreciate this, imagine the consequences of skipping the first letter in the sentence *The dog bit the mailman in the leg* but retaining the same number of letters per word. You'd be left with the nonsensical sentence *Hed ogb itt hem ailmani nt hel eg.* If a ribosome did this while reading genetic code, the garbled message would result in a scrambled protein containing an entirely incorrect amino acid sequence. On top of that, if the garbled message contained one of the three stop codons, the translation process would terminate prematurely. Gene expression would be disrupted.

Herein lies the most basic power of CRISPR—it can destroy a gene's ability to produce a functional protein. If a CRISPR-edited gene ends up with a small insertion or deletion, the corresponding mRNA produced from that gene will be similarly perturbed. And the majority of the time, those extra or missing letters will disrupt the strict three-letter grouping of genetic code, so the protein will be wildly mutated or, more commonly, not produced at all. In any case, the protein can't play its normal role. Geneticists refer to this as a gene knockout, or KO, just like in boxing, since the gene's function has effectively been shut off.

When animal geneticists began using CRISPR, they sought to cre-

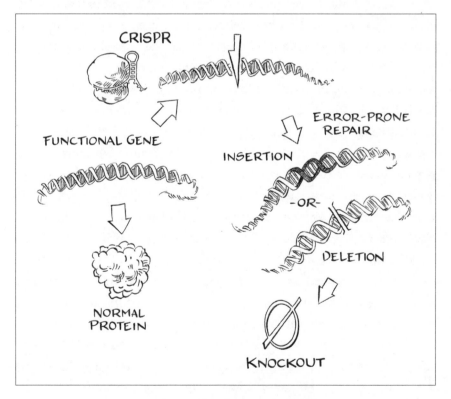

Creating gene knockouts with CRISPR

ate gene knockouts that would manifest themselves in obvious ways. One of the favorite targets was a gene called *TYR*. Having arisen more than half a billion years ago, the *TYR* gene is widely distributed among animals, plants, and fungi; it produces a protein called tyrosinase that is involved in synthesizing melanin, an important pigment. *TYR* mutations in humans lead to a deficiency in tyrosinase and cause type I albinism, a genetic condition associated with vision defects, pale skin lacking pigmentation, and red eyes. If CRISPR was programmed to edit the mouse version of the *TYR* gene, would it cause the mice to have an albino disorder? In 2014, a research team at the University of Texas designed CRISPR to target a twenty-letter DNA sequence in the *TYR* gene and injected it into fertilized eggs. The resulting births were striking; although all of the mouse pups had normal dark-haired parents, many of the pups were perfectly white-haired and red-eyed. Such a result could only be explained by DNA mutations that had scrambled decryption of the *TYR* gene. Changing the skin, hair, and eye color of a mouse was about as clear and profound an effect as you could ask for.

While DNA-sequencing data confirmed that gene editing had occurred, the beauty of targeting the *TYR* gene was that the researchers could check their results visually. Simply counting how many mouse pups were black (the pups in which gene editing had not occurred) and how many were white (the pups in which gene editing had occurred) gave a remarkably accurate measure of CRISPR's efficiency. One could also track how that efficiency changed over time as various laboratories optimized both the design and preparation of CRISPR. In the Texas study, only 11 percent of mice progeny were fully albino, and pictures of the litters revealed a salt-and-pepper pile of infant pups, with quite a bit more pepper than salt. Barely a year later, a Japanese research team repeated the same experiment but with a few minor tweaks and achieved efficiencies of 97 percent, with thirty-nine out of forty pups displaying a radiant, homogeneous albino exterior. In a matter of weeks, the team had permanently and precisely altered the genetic composition of an entire generation of animals (and all their future progeny) in a way that nature had never intended.

Gene knockouts are just one of the many gene-editing strategies that researchers have perfected with CRISPR. Often, genetic engineers need to do better than indiscriminately mutating a gene with random DNA insertions or deletions. After all, a major goal of gene editing, at least in terms of medical application, is to cure genetic diseases, the vast majority of which arise from heritable mutations that inactivate critical genes. In these cases, a gene knockout will be useless, since the patients already suffer from nonfunctional genes. What scientists need is a way to target, edit, and correct single-letter DNA mistakes.

Fortunately, cells possess the machinery to perform a second type of repair, one that is far more precise and controlled than merely gluing broken DNA back together. Instead of joining DNA segments unrelated in sequence, this alternative mode — a pathway that early gene-editing researchers used to their advantage — exclusively rejoins segments that are similar in sequence. This pickiness explains the two synonymous terms that refer to this process: *homologous recombination* and *homology-directed repair*.

Homologous recombination is similar to how a photographer assembles a landscape panorama from three overlapping photos. To get the alignment just right, she has to correctly overlap the outer regions of the middle photo and the inner regions of the outer photos. If the middle part of the panorama is cut or damaged, she can take a duplicate of that middle photo and use the same match-up principle to reconstruct the panorama. And if the real-life landscape changes — say, if a new tower is erected or a large tree falls — she can seamlessly update the panorama by slotting in a new picture using the same approach.

As it turns out, enzymes in the cell carry out analogous kinds of cut-and-paste operations on the panorama that is DNA. The repair option we've already heard about, the error-prone end joining, occurs when cells are confronted with a broken chromosome and haphaz-

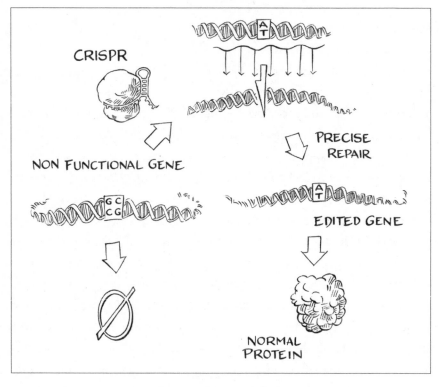

CRISPR

NON FUNCTIONAL GENE

PRECISE REPAIR

EDITED GENE

NORMAL PROTEIN

Homologous recombination with CRISPR

ardly reattach the cut ends, akin to the photographer reassembling a panorama that's missing a sliver of the landscape. But when a cell is faced with a broken chromosome as well as a second piece of DNA that matches the two broken ends — a repair template, much like the photographer's duplicate photo — it chooses the better repair option: it pastes the DNA fragment into the broken chromosome while maintaining a perfect overlap between the matching ends. This strategy means that a harmful genetic mutation at or near the site targeted by CRISPR can be permanently replaced with a healthy new DNA sequence. As long as researchers combine CRISPR with a repair template that matches the area of the broken gene, the cell will gladly grab the replacement part and use it to patch up the damage.

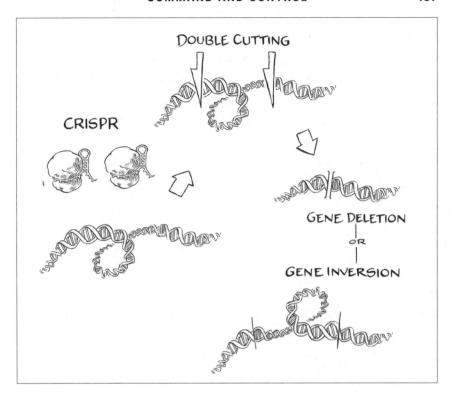

Gene deletion or inversion with CRISPR

In addition to tweaking genes in subtle ways, through error-prone (nonhomologous) or precise (homologous) repair, researchers have employed CRISPR to chop out or flip large swaths of DNA, which allows them to alter a much larger region of the genome. The method takes advantage of the fact that cells will do just about anything to maintain the integrity of a chromosome. By arming Cas9 with two different RNA guides, researchers can program CRISPR to slice apart a chromosome in two adjacent genes; the cell survives the attack by reassembling the chromosome in one of three ways.

With twice as many broken DNA ends for the cell to deal with, option one is for it to kick end-joining repair into high gear and mend both damaged sites concurrently by gluing everything back together. Often, however, the window of opportunity for this repair mode is narrow, due to the constant helter-skelter motion of molecules in the cell. If the segment of DNA between the two cut sites diffuses away, a cell will settle for option two, in which the intervening segment is simply left out and the outermost broken ends are glued back together. This repair mode is comparable to how old-school film editors removed scenes from a movie reel: they simply cut the film in two places — at the beginning and the end of the scene — tossed the scrapped frames, and pasted the new ends together.

The third repair option involves the inversion of the intervening DNA segment. In this case, the cut-out section of DNA is jostled around in such a way that it stays roughly in place but gets flipped around, so its front end is where its back end used to be and vice versa. The same enzymes that facilitate end-joining repair will blindly reattach the missing piece, regardless of its orientation.

There is another category of CRISPR applications that has nothing to do with gene editing. Instead of utilizing CRISPR's DNA-cutting capability, scientists literally break the tool — on purpose. By intentionally disabling the molecular scissors, they can manage the genome from afar — not by permanently editing its DNA, but by changing the way that its DNA gets interpreted, translated, and expressed. Just as invisible strings give a marionettist near complete control over the actions and movements of a puppet, this noncutting version of CRISPR allows scientists to direct the behaviors and outputs of the cell.

The groundwork for this puppeteer function was actually laid early on in my lab's research into CRISPR-Cas9. Proteins are typically composed of hundreds to thousands of amino acid building blocks, a large majority of which are there to define the protein's overall three-dimensional shape; only a few amino acids contribute the critical chemical groups that enable the enzyme to catalyze specific reactions. When Martin Jinek was first characterizing Cas9's biochemical function, he demonstrated exactly which amino acids of the enzyme chemically cleaved, or sliced apart, each strand of the DNA double helix. By mutating those amino acids, he created a version of Cas9 that had completely lost the ability to cut DNA but, remarkably, could still interact with the guide RNA and tightly attach itself to matching DNA sequences. We had broken the catalytic core, but the deactivated Cas9 still retained part of its function — it could hunt down and locate specific DNA sequences in the genome; it just couldn't cut them. Similar research was published by Virginijus Siksnys and colleagues.

Nearby, Stanley Qi, a Berkeley PhD, was starting his own lab at the University of California, San Francisco. Working with Jonathan Weissman and Wendell Lim, both professors at UCSF, Stanley demonstrated that a deactivated version of CRISPR had its own uses for manipulating the genome. Rather than introducing permanent genetic changes by editing DNA, the deactivated CRISPR allowed scientists to make temporary changes that would not alter the underlying genetic information of a cell but nevertheless affected how genetic information was expressed. In particular, he transformed CRISPR into a gene-expression controller that could turn genes on or off or dial them up or down, much like a dimmer adjusts lighting.

The deactivated CRISPR system functions like a molecular packhorse. Instead of targeting specific genes with the end goal of cutting the DNA, scientists combine Cas9 or the guide RNA with protein payloads and then program CRISPR to ferry those payloads to specific genes in the cell. The protein cargo comprises molecules that influence how genes are expressed, either "brightening" or "dimming" their functional output.

Gene-expression control — the complex and overlapping inputs that govern when and for how long genetic information in the form of DNA is turned into protein — is arguably as important to biology as the underlying genetic information itself. Nearly all fifty trillion or so cells in the human body contain the same genome, and yet these highly diverse cell types are of unique shapes and sizes and are arranged into complex organs that have different properties and functions. Some cells attack pathogens in the blood, others expand and contract to pump blood through the body, and still others store memories in the central nervous system; the only thing differentiating immune cells, heart cells, and brain cells is the precise pattern of gene expression that created them. Moreover, the genetic mutations that cause cancer and disease often have their dire effects not because they completely inactivate genes but because they cause genes to express themselves in the wrong way.

The ability to activate or interfere with gene expression is nearly as powerful as the ability to edit the genes themselves. Think of the cell as the largest symphony in the world, made up of more than twenty thousand different instruments. In a healthy, normal-functioning cell, the various symphonic voices are perfectly balanced; in malignant cancer cells or infected cells, the balance is disrupted, with some instruments playing too loud and others too soft. Sometimes, DNA editing is too crude an approach to return the symphony to its normal state — it would be akin to removing or replacing instruments outright. The deactivated CRISPR system offers a way to fine-tune any instrument in the orchestra — that is, any gene in the genome — with greater sensitivity.

Armed with the complete CRISPR toolkit, scientists can now exert nearly complete control over both the composition of the genome and its output. Whether that's done through sloppy end joining or precise homologous recombination, by one cut or multiple cuts or even no cuts at all, the range of possibilities is immense. And the pace of further tool development isn't slowing down. Genetic engineers have built fluorescent versions of CRISPR so that the three-dimensional organization of genes can be visualized inside cells; versions that target mRNA instead of DNA, enabling a unique kind of genetic control; versions that introduce barcodes into the genome, allowing researchers to record a cell's history directly in the language of DNA; and on and on. It often feels like the genome-engineering applications made possible by CRISPR are limited only by our collective imagination. In light of this incredible versatility,

I think it's safe to say that CRISPR will increasingly become a tool of choice for all biologists, regardless of their line of work.

It has been nothing short of exhilarating to see all these incredible technical possibilities come to fruition through the efforts of just a few dozen scientists at first, then hundreds, then even thousands, as more and more researchers adopted CRISPR tools. As any inventor or innovator knows, the feeling of satisfaction when a new invention is embraced by others is unparalleled. Widespread adoption is also the quickest way for a technology to be rapidly refined and reimagined.

The sudden explosion of research into CRISPR was partly the result of its diverse capabilities and partly the result of its incredible range. As the CRISPR toolbox has expanded, no letter of DNA in the genome, no gene or combination of genes, is beyond reach. As I'll explain in the chapters ahead, exploiting this power in humans promises to reshape the treatment of cancer and genetic diseases, and its application in plants and animals provides opportunities to improve food production, eradicate certain pathogens, and even resurrect extinct species. It's no wonder that a few months after the first reports of gene editing with CRISPR were published, *Forbes* magazine predicted that this technology would change biotech forever.

But the real reason that CRISPR exploded onto the biotech scene with such force and vitality was its low cost and ease of use. CRISPR finally made gene editing available to all scientists. Previous tools — primarily ZFNs and TALENs — were difficult to design and prohibitively expensive. For this reason, many labs, including my own, were unwilling to take on the challenges of research using gene editing. With CRISPR, however, scientists can easily design a version to target their gene or genes of interest, prepare the requisite Cas9 protein and guide RNA, and execute the experiments themselves using standard techniques, all within mere days and without requiring any outside assistance. The only thing necessary to get started is a copy of the basic CRISPR-containing artificial chromosome, or plasmid. This need has been conveniently met on a massive scale by the nonprofit organization Addgene, a highly successful and ever-expanding plasmid repository and plasmid-distribution service.

Addgene is like Netflix, only for plasmids. Once Martin and I had submitted our CRISPR article, we sent his plasmids to Addgene for safekeeping, much as film studios license their movies to Netflix. Many other research laboratories that produce CRISPR plasmids do the same. Addgene keeps careful track of the plasmids it has on file, advertises the plasmids and their exact specifications on a website, and generates thousands of duplicate copies that can be distributed to eager customers. The cost to academic laboratories: in 2016, just sixty-five dollars per plasmid. By easing the burden on plasmid producers and satisfying requests from plasmid consumers, Addgene has helped to ensure that any academic or nonprofit lab in the world can obtain research materials, including those needed to employ the CRISPR technology, for its particular experimental needs. In 2015 alone, Addgene shipped some sixty thousand CRISPR-related plasmids to researchers in over eighty different countries.

Computers have also made gene editing easier than ever before. Using advanced algorithms that incorporate all the relevant design principles, including empirical data from the scientific literature on what kinds of targeting sequences work better than others, various software packages offer researchers an automated, one-step method to build the best version of CRISPR to edit a given gene. Far from making scientists lazier, these algorithms have enabled some of the most complex and sophisticated gene-editing experiments to date: the design and execution of genome-wide screens, in which CRISPR is exploited to edit every single gene in the genome.

Today, thanks to these features of CRISPR, an aspiring scientist with the most basic training can accomplish feats that would have been inconceivable just a few years ago. It's become something of an old saw in our young field: what used to require years of work in a sophisticated biology laboratory can now be performed in days by a high school student. Some experts have suggested that, with today's tools, anyone can set up a CRISPR lab for just $2,000. Others predict a rise in do-it-yourself biohackers, eager tech enthusiasts hoping to dabble in CRISPR-based gene editing in their own homes. CRISPR was even the star of a crowdfunded venture that raised well over fifty thousand dollars to generate and distribute DIY gene-editing kits. For $130, donors received "everything you need to make precision genome edits in bacteria at home."

CRISPR has made gene editing available to the masses and is poised to turn this once-esoteric practice into a hobby or a craft, just like homebrewing beer. (In fact, editing the yeast genome to make new flavors of beer is another one of the many interesting and unexpected uses of CRISPR that I have come across.) In many ways, this is exciting — but there's also something unsettling about the rapid spread of this powerful tool.

The democratization of CRISPR will accelerate the process of research and development that I've described in this chapter, but it will also lead to uses of this technology that people are not yet prepared for — and whose effects can't be contained within the lab. Scientists the world over have already begun using CRISPR on other species in ways that defy imagination, and it won't be long before human genomes are given the same treatment.

How will we even begin to weigh the costs and benefits of tampering with our own genetic code? Will we be able to agree on the right way to use CRISPR, and will we be able to prevent it from being abused?

With our mastery over the code of life comes a level of responsibility for which we, as individuals and as a species, are woefully unprepared. In the next part of this book, I will explore some of the dilemmas that have arisen as a result of the CRISPR revolution, as well as the incredible opportunities for good that have come out of it. Weighing the dangers inherent in a technology like CRISPR against the responsibility to use its power for the benefit of humanity and our planet will be a test like no other. Yet it's one that we must pass. Given the stakes, we simply have no alternative.

Part II
THE TASK

5

THE CRISPR MENAGERIE

TOMATOES THAT CAN SIT in the pantry slowly ripening for months without rotting. Plants that can better weather climate change. Mosquitoes that are unable to transmit malaria. Ultra-muscular dogs that make fearsome partners for police and soldiers. Cows that no longer grow horns.

These organisms might sound far-fetched, but in fact, they already exist, thanks to gene editing. And they're only the beginning. As I write this, the world around us is being revolutionized by CRISPR, whether we're ready for it or not. Within the next few years, this new biotechnology will give us higher-yielding crops, healthier livestock, and more nutritious foods. Within a few decades, we might well have genetically engineered pigs that can serve as human organ donors — but we could also have woolly mammoths, winged lizards, and unicorns. No, I am not kidding.

It amazes me to realize that we are on the cusp of a new era in the history of life on earth — an age in which humans exercise an unprecedented level of control over the genetic composition of the species that co-inhabit our planet. It won't be long before CRISPR allows us to bend nature to our will in the way that humans have dreamed of since prehistory. When that will is directed toward something constructive, the results could be fantastic — but they might also have unintentional or even calamitous consequences.

The impact of gene-edited plants and animals is already being felt in the scientific community. For example, researchers have harnessed CRISPR to generate animal models of human disease with far greater precision and flexibility than before — not just in mice, but in whatever animals best exhibit the disease of interest, whether it be monkeys for autism, pigs for Parkinson's, or ferrets for influenza. One of the most interesting aspects of the CRISPR technology is the way it enables the study of features unique to certain organisms, such as limb regeneration in Mexican salamanders, aging in killifish, and skeletal development in crustaceans. I love the notes and pictures colleagues send me describing their CRISPR experiments — the beautiful butterfly-wing patterns whose genetic underpinnings they've uncovered, or the infectious yeast whose ability to invade human tissues they've dissected at the level of individual genes. These kinds of experiments reveal new truths about the natural world and about the genetic similarities that bind all organisms together. They're enormously exciting to me.

At the other end of the spectrum are gene-editing applications that read more like science fiction than the contents of a scientific journal. For example, I was amazed to learn that several research teams are using CRISPR to "humanize" various genes in pigs in the hope that life-threatening organ-donor shortages might one day be solved by xenotransplantation — the transfer of organs grown in pigs (or other animals) into human recipients. In a sign of the kinds of aesthetic changes to animals that are now possible, companies have used gene-editing technologies to create new designer pets, such as gene-edited micropigs that never grow larger than small dogs. And in a page taken straight out of a famous book-to-film sci-fi franchise, some laboratories are pursuing a venture known as de-extinction, which is nothing less than the resurrection of extinct species through cloning or genetic engineering. My friend Beth Shapiro, a professor at the University of California, Santa Cruz, is excited to use this strategy to re-create extinct species of birds for the purpose of studying their relationships to modern species. Along the same lines, efforts are already under way to convert the elephant genome into the woolly mammoth genome, bit by bit, using CRISPR.

Ironically, CRISPR might also enable the opposite: forcible extinction of unwanted animals or pathogens. Yes, someday soon, CRISPR might be employed to destroy entire species — an application I never could have imagined when my lab first entered the fledgling field of bacterial adaptive immune systems just ten years ago.

Some of the efforts in these and other areas of the natural world have tremendous potential for improving human health and well-being. Others are frivolous, whimsical, or even downright dangerous. And I have become increasingly aware of the need to understand the risks of gene editing, especially in light of its accelerating use.

CRISPR gives us the power to radically and irreversibly alter the biosphere that we inhabit by providing a way to rewrite the very molecules of life any way we wish. At the moment, I don't think there is nearly enough discussion of the possibilities it presents — for good, but also for ill. It's a thrilling moment in the life sciences, but we can't let ourselves get carried away. It's important to remember that, while CRISPR has enormous and undeniable potential to improve our world, tinkering with the genetic underpinnings of our ecosystem could also have unintended consequences. We have a responsibility to consider the ramifications in advance and to engage in a global, public, and inclusive conversation about how to best harness gene editing in the natural world, before it's too late.

In 2004, a team of European scientists solved a long-standing mystery facing barley breeders. The researchers had discovered gene mutations that made the plant resistant to a pernicious fungus that causes a disease known as powdery mildew — a blight that had long tormented farmers of the elite barley cultivars grown throughout Europe. The mutated, fungus-resistant barley strain could be traced to barley seeds collected from granaries in southwestern Ethiopia during German expeditions in the late 1930s. There, sometime after the domestication of barley (around ten thousand years ago), a mutated version of a gene called *Mlo* appeared spontaneously and was selected for by farmers eager to cultivate only the healthiest-looking and highest-yielding plants.

This human-influenced evolutionary process — natural mutation followed by artificial selection rather than natural selection — is how agriculture has developed for millennia. As pioneering agriculturist Luther Burbank remarked in a speech in 1901, species weren't fixed and unchangeable but rather "as plastic in our hands as clay in the hands of the potter or colors on the artist's canvas, and can readily be molded into more beautiful forms and colors than any painter or sculptor can ever hope to bring forth." In fact, the discovery of the protective *Mlo* gene mutation in barley originated with a German cultivar that had been irradiated with x-rays in 1942. Scientists had found that exposing seeds to radiation (x-rays or gamma rays, for example) or bathing them in mutation-inducing chemicals peppered the genome with sporadic new mutations from which plants with desirable traits could be bred.

The mutant strains produced using these methods are genetically altered in unknown ways across hundreds or even thousands of different genes. If among those random genetic alterations, the strains happen to share similar mutations, such as in the *Mlo* gene, the resulting plants may all have the trait that was desired — in this case, fungus-resistant barley. A decade after the 2004 identification of the protective *Mlo* mutation in barley, disruption of this same gene was linked to powdery mildew resistance in several other plants. This raised the exciting possibility that many more crops could be endowed with powdery mildew resistance by altering the *Mlo* gene.

Herein lies the promise of gene editing. Compared to conventional breeding methods — including natural mutagenesis, induced mutagenesis using x-rays or chemicals, and hybridization between different plant species (which floods the genome with thousands of new genes) —

CRISPR and its kindred technologies give scientists a level of control over the genome that is unparalleled. The possibilities of this technology for agriculture were highlighted in my mind when, in 2014, scientists at the Chinese Academy of Sciences used gene-editing tools, including CRISPR, to alter the six copies of the *Mlo* gene in *Triticum aestivum,* or bread wheat, one of the world's most important staple crops. Plants that had all six mutated *Mlo* genes were resistant to powdery mildew, a fantastic result, and furthermore, the researchers didn't have to worry about harmful or undesired effects of any other mutations because only the *Mlo* genes had been edited. Whether the desired changes are gene knockouts (as in *Mlo*), gene corrections, gene insertions, or gene deletions, scientists can alter the genome with unprecedented, single-letter accuracy, and do so for virtually any gene and any DNA sequence.

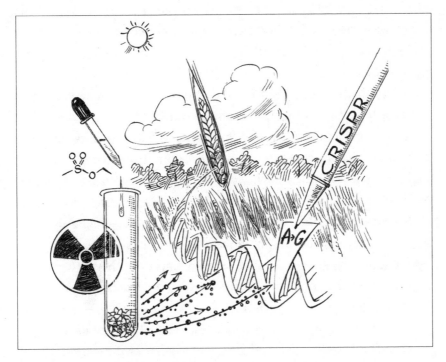

Methods of introducing DNA mutations in plants

Powdery mildew is just one example of the agricultural challenges that can be addressed with CRISPR. In the few years since its inception, CRISPR has been harnessed to edit genes in rice that confer protection against bacterial blight; to endow corn, soybeans, and potatoes with natural resistance to herbicides; and to produce mushrooms that are impervious to browning and premature spoiling. Scientists have used CRISPR to edit the genome of sweet oranges, and a team of California researchers is now attempting to apply the technology to save the U.S. citrus industry from a bacterial plant disease called huanglongbing — a Chinese name that translates as "yellow dragon disease" — that has devastated parts of Asia and now threatens orchards in Florida, Texas, and California. In South Korea, scientist Jin-Soo Kim and his colleagues hope gene editing in bananas can help save the prized Cavendish variety from extinction, an outcome threatened by the spread of a devastating soil fungus. And elsewhere, researchers are even toying with the possibility of inserting the entire bacterial CRISPR system, reprogrammed to slice up plant viruses, into crops, providing them with a completely new antiviral immune system.

I'm particularly intrigued by opportunities to use gene editing to produce healthier foods. Two examples stand out. The first involves soybeans, which provide around fifty million tons of soybean oil annually. Unfortunately, soybean oil contains unhealthy levels of trans fats that have been linked to high cholesterol and heart disease. Recently, food scientists at a Minnesota company called Calyxt used the TALEN gene-editing technology to alter two soybean genes, generating seeds with a drastic reduction in the unhealthy fatty acids and an overall fat profile similar to that of olive oil. They accomplished this without causing any unintended mutations and without introducing any foreign DNA into the genome.

The second example involves potatoes, the world's third most important food crop after wheat and rice. The lengthy cold storage required to increase potatoes' shelf life can lead to cold-induced sweetening, a phenomenon in which starches are converted into sugars such as glu-cose and fructose. Any cooking process involving high heat — neces-sary to make french fries and potato chips — converts these sugars into acrylamide, a chemical that is a neurotoxin and a potential carcinogen. Cold-induced sweetening also causes potato chips to brown and take on a bitter taste, which results in a huge amount of waste; processing plants discard 15 percent of their potatoes a year for this reason. Using gene editing, researchers at Calyxt easily addressed the problem in Ranger Russet potatoes: they inactivated the single gene that produced glucose and fructose. The result: a 70 percent drop in acrylamide levels in potato chips made with the enhanced spuds, and no chip browning.

Food scientists are ecstatic over the possibilities of easy gene editing. But there's still a big elephant in the room: Will producers and consumers embrace precision gene-edited crops the same way they have embraced the thousands of crops whose genomes have been randomly mutated with x-rays, gamma rays, and chemical mutagens? Or will gene-edited crops suffer the same fate as GMOs, another type of genetically altered food and one that has met with incredible and, I would argue, misin-formed opposition despite its vast potential for good?

As the CRISPR technology has spread around the world, food politics is one of the many areas in which I've had to educate myself. Knowing that gene-edited plants and animals will inevitably be compared to GMOs, I've specifically committed myself to learning what different national governments and public-interest groups even mean when they use the term *genetically modified organism.*

The U.S. Department of Agriculture (USDA) defines genetic modi-fication as "the production of heritable improvements in plants or ani-mals for specific uses, via either genetic engineering or other more tra-ditional methods." This broad umbrella could cover newer technologies like gene editing as well as also older methods like mutation breeding. Indeed, under this definition, just about every food we eat, aside from

wild mushrooms, wild berries, wild game, and wild fish, could be considered a GMO.

A more common definition of GMO, however, includes only those
organisms whose genetic material has been altered using recombinant
DNA technology and so-called gene splicing, in which foreign DNA
sequences are integrated into the genome. Since 1994, when the first
commercially grown GMO plant approved for human consumption was
introduced — a slow-ripening tomato variant known as the Flavr Savr —
well over fifty GMO crops have been developed and approved for commercial cultivation in the United States, among them canola, corn, cotton, papaya, rice, soybean, squash, and many more. In 2015, 92 percent
of all corn, 94 percent of all cotton, and 94 percent of all soybeans grown
in the United States were genetically engineered in this way.

The altered crops offer considerable environmental and economic advantages. By planting crops that have enhanced abilities to protect themselves against pests, farmers can attain higher yields while reducing their
reliance on harsh chemical pesticides and herbicides. Genetic engineering has also saved entire industries, such as the Hawaiian papaya, from
the scourge of viruses, and it may soon prove critical for protecting other
fruits, like bananas and plums, that are threatened by newly emerging
pathogens.

Despite these benefits, and despite the fact that hundreds of millions
of people have consumed GMO foods without any issues, these foods
remain the target of vociferous criticism, intense public scrutiny, and
strident protest, most of it without merit. Rallying cries have centered
on a small handful of studies that claimed to reveal adverse effects on
consumer health or the environment — for example, stating that GMO
potatoes gave rats cancer and that GMO corn killed monarch butterflies
— but these reports have been discounted in numerous follow-up studies and condemned by the broader scientific community. In fact, GMOs
have been subjected to some of the most careful regulatory review of
any human consumables on the market, and there is near-unanimous

consensus that GM food is every bit as safe as conventionally produced food. GMOs have received support from federal regulators in the United States, the American Medical Association, the U.S. National Academy of Sciences, the Royal Society of Medicine in the UK, the European Commission, and the World Health Organization. Nevertheless, nearly 60 percent of Americans perceive GMOs as unsafe.

The disjunction between scientific consensus and public opinion on the topic of GMOs is disturbing, to say the least. I see it as partly a reflection of the breakdown in communication between scientists and the public at large. Already in my relatively short time working on CRISPR, I've discovered how challenging it can be to maintain a constructive, open dialogue between these two worlds — but also how necessary that kind of communication is for the advancement of scientific discoveries.

The perception that GMOs are somehow unnatural and perverse is a case in point. Almost everything we eat has been altered by humans, often by generating random mutations in the DNA of seeds used to breed plants with desired traits. Thus, the distinction between "natural" and "unnatural" has been obscured. Red grapefruits created by neutron radiation, seedless watermelons produced with the chemical compound colchicine, apple orchards in which every tree is a perfect genetic clone of its neighbors — none of these aspects of modern agriculture is natural. Yet most of us eat these foods without complaint.

CRISPR and related gene-editing technologies will further complicate the debate over genetically modified foods by blurring the lines between GMO and non-GMO products. Conventional GMOs contain foreign genes randomly inserted into the genome; these genes produce novel proteins that give the organism a beneficial trait it did not previously possess. Gene-edited organisms, by contrast, contain tiny alterations to existing genes that give the organism a beneficial trait by tweaking the levels of proteins that were already there to begin with — without adding any foreign DNA. In this respect, gene-edited organisms are often no different than those organisms produced by mutation-inducing chemi-

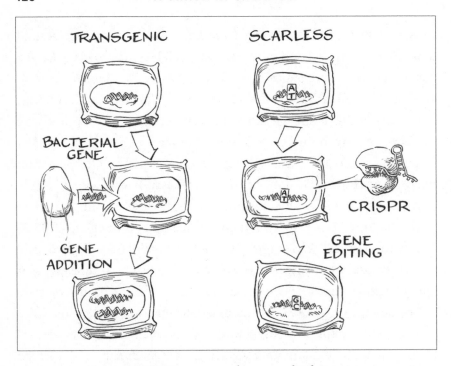

Transgenic GMOs versus scarless gene-edited organisms

cals and radiation. Furthermore, scientists have used methods to avoid leaving any traces of CRISPR in the plant genome once the gene-editing task is complete. For example, CRISPR molecules can be manufactured, purified, and assembled in the laboratory (just as we demonstrated in our 2012 article), and then delivered to plant cells in this fast-action formulation so they go to work on the genome immediately. In a matter of hours, Cas9 and the guide RNA will edit the gene of interest and then be degraded by natural recycling processes in the cell. I'm hopeful that, in time, this kind of scarless gene editing will help win public acceptance for crops and other plants improved by these precision methods.

Controversy is slowly brewing around gene-edited organisms, however. Some of the first activist-led protests over the new technology took place in the spring of 2016. CRISPR researchers have even been threatened by activists who had previously focused their attention on GMOs.

One of the biggest challenges facing agricultural companies, farmers, consumers, and especially government officials is how to classify and regulate gene-edited crops. Many scientists classify them as products of new breeding techniques, or NBTs, whereas protesters feel that the gene-edited crops are nothing but hidden GMOs and that scientists are trying to sneak them into grocery stores through the back door. In many ways, the issue boils down to product versus process: Should regulations for a newly generated crop consider only the final product, or should they also take into account the process that was employed to develop it? To return to the powdery mildew example, does it matter that an advanced form of gene editing was used to make bread wheat resistant to this disease, even if the resulting strain of wheat is no different than what could have theoretically been generated by natural or induced mutations?

At the moment, new genetically modified crops face a confusing array of regulatory hurdles, with jurisdiction split among the Food and Drug Administration, the Environmental Protection Agency, and the U.S. Department of Agriculture. The approval process is both long and expensive and includes what some consider to be an unfair and onerous set of requirements. Many smaller companies are locked out of the GMO field altogether because of prohibitive costs, allowing big agribusinesses to monopolize the market. I was surprised to discover that even academic scientists have difficulty studying genetically modified crops in field tests because of burdensome restrictions.

Thankfully, this situation is beginning to change. The USDA has quietly started informing companies that the new generation of gene-edited crops will not require USDA approval — although they must still be approved by the FDA. Herbicide-resistant canola generated by gene editing has been approved for use in Canada and was deemed not to fall under the purview of the USDA. Similarly, the gene-edited soybeans and potatoes produced by Calyxt scientists using TALENs skipped past USDA regulation, as did some thirty other types of genetically modified plants.

And although CRISPR is relatively new to the scene, DuPont Pioneer predicts that CRISPR-based plant products will be on the market by the end of the decade.

Meanwhile, in 2015, the White House Office of Science and Technology Policy announced that it would revisit the regulation of genetically engineered crops and animals in light of new technology developments and the fact that current policy hadn't been updated since 1992. The way that genetically engineered products are marketed is also in flux, with the 2016 passage of federal legislation that requires labeling of foods containing genetically modified ingredients.

Changes in regulation such as these are important, but unless public attitudes toward genetically enhanced foods change with them, we as a society won't be able to benefit from the full potential of CRISPR. Biotechnology can help us shore up our food security, stave off malnutrition, adapt to climate change, and prevent environmental degradation around the world. This progress will remain out of reach, however, until scientists, companies, governments, and the public at large work together to make it happen. Each of us can contribute to this partnership in a very basic way. It starts with an open mind.

Agribusiness isn't interested in CRISPR for crops alone; livestock, too, will be widely gene edited in the near future. Yet given the formidable obstacles that GMO plants have encountered, gene-edited animals are likely to face many of the same regulatory hurdles and even stronger opposition. Here, as on that front, we have much to gain and possibly even more to lose.

The first genetically engineered animal to be approved for human consumption in the United States — a fast-growing GMO salmon breed, called AquAdvantage — made it to market only after a twenty-year battle with FDA regulators and at a cost of over eighty million dollars for its developer. The gene-spliced salmon contains an extra growth hormone gene, resulting in a fish that reaches market weight in half the time of a conventionally farmed salmon and without any changes to its nutritional content or any increased health risks for either the fish or the humans who eat it. Advocates argue that high-yielding farmed salmon would be a boon to the environment because they would reduce depletion of wild fish stocks, decrease the amount of salmon imported into the United States (currently 95 percent), and deliver fish to the market with a carbon footprint that is around twenty-five times less than for conventional salmon. Still, as with GMO crops, the backlash against genetically modified salmon has been intense; opponents have branded the animals "Frankenfish" and claimed that the salmon endanger consumers' personal health as well as wild fish ecosystems. A 2013 *New York Times* poll found that 75 percent of respondents wouldn't eat GMO fish, and consumer criticism has led more than sixty grocery-store chains across the United States — including retail giants such as Whole Foods, Safeway, Target, and Trader Joe's — to promise not to sell the salmon.

The AquAdvantage salmon is not the first genetically modified animal that scientists have created for human consumption. In the early 2000s, a Japanese team bred pigs containing a spinach gene that altered the way the animals metabolized fatty acids; the transgenic swine had a healthier fat profile, but the scientists' work was widely condemned, and the pigs never made it out of the lab. Around the same time, a Canadian team created the Enviropig, an environmentally friendly transgenic pig containing an *E. coli* gene that allowed the animals to better digest a phosphorus-containing compound called phytate. Normal pig manure retains high phosphorus levels that leach into streams and rivers, causing algal blooms, the death of aquatic animals, and the production of greenhouse gases; Enviropig manure contained 75 percent less phosphorus, which could have been an enormous benefit to the planet and to the people who lived and worked near pig farms. Despite this, though, and despite reassuring safety data, consumers decried the Enviropig, causing the project's financial backers to pull the plug. The new breed was finally euthanized in 2012.

Against the backdrop of cases like these, the outlook for other genetically modified animals seems bleak. But then again, that all depends on how *genetically modified* is defined by regulators and by the public. The AquAdvantage salmon was endowed with a growth hormone gene from Chinook salmon as well as a short piece of DNA from ocean pout to keep the growth hormone gene switched on. What if, instead, scientists had somehow managed to edit the salmon's genome to ramp up production of its own growth hormone gene without adding any foreign DNA? Would consumers and regulators still consider the salmon a GMO?

This question is sure to arise in the near future, given the rapid pace of research and development into gene-edited livestock. The first engineered animals have already been generated in the lab, and it's only a matter of time before they show up on regulators' doorsteps. Like the AquAdvantage salmon, some of these pioneering animals will have genetic modifications that encourage their growth. But unlike the salmon, they won't just grow faster; they'll also grow bigger.

Using the new powers of precision gene editing — from CRISPR and related technologies — scientists have created gene-edited cows, pigs, sheep, and goats that are stronger and more muscular than average, with striking bodybuilder-like physiques, a trait commonly referred to as double muscling. Far from being some freakish attribute invented in the lab, this mutation is inspired by nature, just like barley's powdery mildew–resistance trait.

Cattle farmers have known about double muscling for years because of its high frequency in two popular cattle breeds: the Belgian Blue and the Piedmontese. These cows have 20 percent more muscle on average, a higher meat-to-bone ratio, less fat, and a higher percentage of desirable cuts of meat, making them a beef producer's dream. In 1997, three labs determined that a single gene was responsible for this exceptional form of muscle development. The gene, called *myostatin,* behaves like a natural brake on the body's production of muscle tissue. The two breeds of cattle these labs were studying have different kinds of mutations — Bel-

Items on Loan

Library name: Suffolk Library
User name: Miss Siobhan O'Connor

Author: Doudna, Jennifer A.,
Title: A crack in creation : the new power to control ev
Item ID: C902233401
Date due: 27/7/2022,23:59
Date charged: 6/7/2022, 16.55

LibrariesNi
****

Make your life easier
****

Email notifications are sent two days before item due dates
Ask staff to sign up for email

gian Blue cows are missing eleven letters of DNA, whereas Piedmontese cows have just a single-letter mutation — but in both cases, the protein product of the *myostatin* gene is defective. In a sense, nature had mirrored previous genetics experiments conducted in mice, where *myostatin* gene knockouts produced similarly brawny creatures that weighed two to three times above average — a gain due exclusively to muscle mass, not fat.

Cows aren't the only animals to exhibit the natural double-muscling trait. Texel sheep, a popular Dutch breed prized for its lean meat and heavily muscled build, also contain a *myostatin* mutation. So do whippets, a dog breed descended from greyhounds that is frequently used in racing, as the dogs have not only the highest running speed of breeds of their weight but also the fastest acceleration of any dog in the world. Whippets of the "bully" variety have broad chests and massive leg and neck musculature caused by the two missing DNA letters in the *myostatin* gene. Other whippets have a normal *myostatin* gene, while yet others are known as heterozygotes because they possess both a normal and a mutated copy of the two parental chromosomes. (A study from the National Institutes of Health found that the fastest whippets are in fact the heterozygotes because they have some extra muscle, but not too much — a kind of genetic Goldilocks.)

Even some humans exhibit the equivalent of double muscling. In 2004, a team of physicians from Berlin published a remarkable study describing a boy who was extraordinarily muscular at birth, with bulging thigh and upper-arm muscles. The child continued to develop abnormally pronounced muscles through age four and could perform incredible feats of strength, like extending his arms while holding a three-kilogram dumbbell in each hand. Given his condition's resemblance to the double muscling in cows and mice, and given the history of unusual strength in his family, the physicians suspected that genetics could explain his physique. After some molecular detective work, they found that both copies of his *myostatin* gene contained knockout mutations and that his mother, a former professional athlete, was a heterozygote with just one mutated gene. Although the prevalence of muscle hypertrophy — as doctors call the double-muscling condition — is extremely rare in humans, at least one other case has since been reported, this one in a Michigan family.

Researchers are now investigating whether replicating this condition with deliberate mutations — that is, stimulating muscle growth by inactivating the *myostatin* gene — may be a viable therapy for treating muscle-wasting diseases such as muscular dystrophy. Some writers have even begun fantasizing about editing the *myostatin* gene in normal individuals to unleash enhanced, superhuman strength — although, as I'll

Double-muscled animals, both natural and created with CRISPR

discuss in the coming chapters, I think that the implications of this sort of nonessential gene editing in humans are troubling.

In livestock, unlike in humans, there are reasons to use gene editing to create new varieties of organisms with advantageous traits. For one thing, small enhancements to animal genomes might result in significant increases in food production. Scientists have already employed gene editing to develop new breeds of double-muscled cows, sheep, pigs, goats, and rabbits. It's not hard to imagine what animals like these could mean for human nutrition if they were made available to farmers. High yields of lean meat, together with low body fat, have been a major goal of breeding in the livestock industry, and gene editing offers an easy way to achieve it. In one report, gene-edited pigs had over 10 percent more lean meat than their unedited counterparts, as well as a substantial decrease in total body fat and increased meat tenderness. At the same time, the meat's nutritional content and the animals' development, diet, and overall health were unaffected. Because the tweaked porcine genome contains no traces of transgenes, producers hope that the pigs will be regulated no differently than animals like Belgian Blue cattle, which developed double muscling through natural mutations.

Since CRISPR makes it easy to edit multiple genes, numerous new traits can be introduced simultaneously. For example, Chinese scientists working with goats targeted the *myostatin* gene as well as a growth factor gene known to control hair length. In humans, naturally occurring mutations in the growth factor gene cause a condition characterized by excessively long eyelashes, and the mutations have been linked to hair length in cats, dogs, and even donkeys. The scientists performed gene editing in a breed of goats known as Shannbei, cultivated for both its desirable meat and its hair fibers, which are used to produce fine cashmere. The scientists injected 862 embryos and transferred 416 into recipient mothers; 93 kids were born, 10 of which contained mutations in both genes. The enhanced goats can now serve as the starting point for new breeds that provide higher yields of both food and cashmere for their farmers.

Other scientists are using gene-editing tools to bias reproduction so that chickens produce only females (on egg farms, male chicks are typically culled within a day of hatching), farmed fish are sterile (and can't pollute natural stocks), and beef cattle produce only the profitable males (since females convert feed to muscle far less efficiently). Cattle genomes are being altered to resist a parasite known to cause sleeping sickness, and porcine genomes are being modified so that pigs can be fattened with less food. In Australia, a team is trying to alter a gene in chickens that produces one of the most common allergenic proteins in chicken eggs, and similar strategies have been proposed to remove allergens in cow milk.

Animals' genes can also be edited to make the creatures healthier and more disease-resistant, as recent experiments in pigs have convincingly demonstrated. One of the major diseases facing the swine industry is caused by a virus known as PRRSV, which, after being recognized in the United States in the late 1980s, spread rapidly through North America, Europe, and Asia. The virus costs U.S. pork producers more than $500 million annually and reduces production by 15 percent, and a heavy price is paid by the animals themselves; infected pigs suffer from a range of symptoms including anorexia, fever, an increased frequency of aborted and mummified piglets, and severe respiratory problems. Vaccination programs have thus far been unsuccessful, leaving heavier doses of in-feed antibiotics, to stave off secondary bacterial infections, as one of the few options available to treat the animals.

Inspired by theories that it was a particular porcine gene, *CD163*, that allowed viruses to hijack pig cells, a team at the University of Missouri sought to generate virus-resistant pigs by inactivating the problem gene (a strategy not unlike changing the locks of a house to thwart a would-be burglar with a stolen key). After using CRISPR to create gene-knockout pigs, the Missouri researchers sent the animals over to

Kansas State University, along with unedited piglets as controls, to be tested for virus susceptibility. There, the pigs were exposed to some one hundred thousand viral particles and continually monitored. Remarkably, the gene-edited pigs remained completely healthy and free of any traces of virus.

This strategy — saving pigs from viruses by knocking out the genes that viruses depend on — has been so effective that it's already being adopted by other researchers to reduce suffering and waste in other corners of the meat industry. For example, a group of UK scientists has scored a similar victory in the fight against a different virus. The disease it causes, known as African swine fever, also infects domesticated pigs; like PRRSV, the virus is highly contagious and has no vaccines. But the African virus is even more deadly, with some strains causing near 100 percent mortality, typically by intense hemorrhaging within as little as one week of infection. Sadly, this isn't the only way that the disease claims animals' lives; as the virus has swept through Eastern Europe, farmers have resorted to slaughtering pigs, entire herds in some cases, as a last-ditch effort to stop the outbreak.

Noticing that species of pigs native to Africa, including warthogs, appeared to be unaffected by the virus, the UK team zeroed in on a single gene that seemed to explain their remarkable resistance. The warthog version of the gene differs from that of domestic pigs by just a few letters, so the scientists simply edited the domestic pigs' genes to match it without altering any other parts of the genome. Time will tell if the edited pigs possess the same immunity as warthogs and, perhaps more important, if the public will embrace the new genetically modified animals. The researchers, at least, are confident that consumers won't take issue with a tiny refinement that ultimately produces healthier animals, especially an alteration that already exists in nature.

Another example of livestock gene editing comes from a Minnesota company called Recombinetics, which achieved the remarkable feat of genetically modifying cows to prevent them from growing horns. The

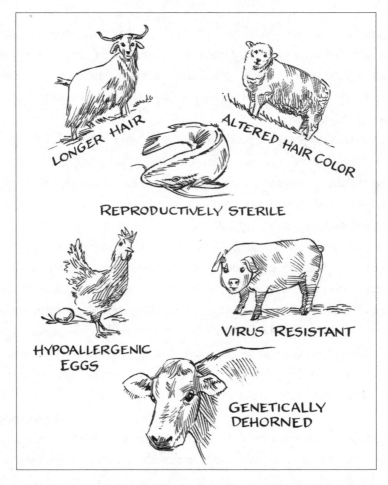

Other gene-edited animals on the horizon

company's goal was to obviate the cruel but widespread practice of cattle dehorning, a common procedure in the U.S. and European dairy industries. Horns make handling confined animals dangerous for farm workers and can also pose a risk to the cows themselves. Food producers typically remove horns at a young age by burning off horn buds with a heated iron, causing tissue damage and a significant amount of stress and pain for the traumatized calves. In the United States alone, well over thirteen million calves are dehorned every year.

Not all cows have horns, though. In fact, many beef cattle breeds — including the popular Angus — are naturally horn-free. In 2012, a German research team discovered the exact genetic cause: a complex mutation involving the deletion of 10 DNA letters and the insertion of 212 DNA letters on chromosome 1. Inspired by this knowledge, the scientists at Recombinetics used gene editing to copy the exact same change into the genome of blue-ribbon dairy bulls, creating cattle whose prized genetics — crafted over centuries of selective breeding for optimal milk production — weren't otherwise altered. The first such animals to be born, two hornless dairy calves named Spotigy and Buri, will never know the horror of having their horn buds removed.

Moving forward, regulators and consumers contemplating such gene-edited livestock will need to decide which matters more: the end or the means; the product or the process that creates it. The hornless cattle might have been produced by years of conventional breeding. Gene editing merely allowed the same outcome to be achieved much more efficiently. If CRISPR and related technologies can eliminate inhumane practices like dehorning, reduce antibiotic usage, and protect livestock from deadly infections, can we afford not to use them?

Breeders and food scientists are not the only ones editing the genomes of animals. So are biomedical scientists, men and women whose goal is to improve the lives of people using methods that have been tested — and in some cases derived — from gene-edited animals.

Animal research is indispensable to the study of human disease, whether it's used to confirm the genetic causes of certain disorders, to evaluate potential drugs, or to test the efficacy of medical interventions like surgery or cell therapy. A crucial starting point is having a robust genetic model — an animal whose condition closely mimics that of the diseased patient group in terms of both physical manifestations and the underlying genetic causes. CRISPR offers an effective, streamlined approach to accomplish this.

The preferred mammalian model organism for biomedical research since the early twentieth century has been the common house mouse (*Mus musculus*), which shares 99 percent of its genes with humans. In addition to their genetic relatedness to us, mice offer other clear advantages. Mice and humans exhibit similar physiological features, such as immune, nervous, cardiovascular, musculoskeletal, and other systems. Mice can be bred in captivity and are easy and cheap to maintain because of their small size, docility, and fecundity. Their accelerated lifespan — one mouse year equals roughly thirty human years — means that the entire life cycle can be studied in just a few years in the lab. And perhaps most important, mice can be genetically manipulated using a variety of approaches — CRISPR being the most powerful and most recent — to mimic a large number of human diseases and conditions. Millions of mice are bred and shipped each year to researchers worldwide, and there are well over thirty thousand unique mouse strains in existence that are used to study everything from cancer and heart disease to blindness and osteoporosis.

But as models, mice also have limitations, because for many human diseases — cystic fibrosis, Parkinson's, Alzheimer's, and Huntington's, among others — they do not exhibit the hallmark symptoms or they have atypical responses to potential treatments. These shortcomings have created a gap in the bench-to-bedside approach of translating research discoveries in the lab to medical treatments in the clinic.

CRISPR will help bridge this gap by making disease modeling in other animals virtually as accessible as it has been in the mouse. This development can already be seen in nonhuman primates. Transgenic monkeys were first generated in the early 2000s, when researchers used viruses to splice foreign genes into the monkeys' genomes, but gene-edited monkeys had never been achieved in the pre-CRISPR era. That changed in early 2014, when a Chinese team created gene-edited cynomolgus monkeys by injecting CRISPR into one-cell embryos, much like the method used in mice one year earlier. In this study, the scientists programmed

CRISPR to simultaneously target two genes: one associated with severe combined immunodeficiency in humans, the other associated with obesity, both of which have clear implications for human health. Since then, other researchers have created cynomolgus monkeys with changes in a gene that is mutated in over 50 percent of human cancers, and rhesus monkeys that carry the mutations that cause Duchenne muscular dystrophy. Gene editing is also being exploited to target genes implicated in neural disorders, taking advantage of the fact that monkey models are uniquely suited for the study of human behavioral and cognitive abnormalities. Although on one level, I feel uneasy about using monkeys in this way, I am also sensitive to the intense need to develop treatments and cures for human disease to alleviate human suffering. These gene-edited monkeys can serve as reliable stand-ins for human patients, allowing scientists to hunt for disease cures without endangering human lives.

The pig has become another popular animal model of human disease, thanks to CRISPR. This is because of its anatomic similarity to humans, its relatively short gestation period, and its large litter size. With suitable guidelines, I view the use of farm animals for biomedical research purposes as more acceptable than the use of companion animals like primates. Indeed, gene-edited pigs have already been used to model pigmentation defects, deafness syndromes, Parkinson's disease, and immunological disorders, and the list will continue to grow.

Many scientists see the pig itself as a potential source of medicine. Someday soon, we might be using pigs as bioreactors to produce valuable drugs like therapeutic human proteins, which are too complex to synthesize from scratch and can only be produced in living cells. Scientists have already been looking to other transgenic animals to produce these biopharmaceutical drugs, or farmaceuticals, as they're colloquially called. The first such drug to be approved by the FDA is an anticoagulant called antithrombin, and it is secreted in the milk of genetically modified goats. Another approved drug is isolated from the milk of transgenic rabbits, and in 2015, the FDA gave the go-ahead for a protein-based drug that is purified from the egg whites of transgenic chickens.

There are numerous benefits to extracting the drugs from transgenic animals rather than from cultured cells, including higher yields, easier scale-up, and lower costs. CRISPR promises to further improve farmaceutical production by giving scientists far better genetic control over creation of the transgenic animals in the first place. For example, experiments in pigs have shown that CRISPR enables outright replacement of pig genes with their human gene counterparts, allowing for more efficient recovery of the therapeutic gene-encoded proteins. When you consider that many of the world's bestselling drugs are protein-based, it's clear that the potential for gene editing in this particular subfield of medicine is enormous.

Some scientists hope that pigs can offer even more: a vast, renewable source of whole organs for xenotransplantation into human recipients. It's not a new idea; pigs have long been considered for this role for some of the same reasons they're favored as disease models — they're easy to breed and reproduce quickly — and the fact that their organs are re-markably comparable in size to those of humans. But for just as long, this dream seemed unrealizable. An array of immunological defenses in the human body make rejection of donor organs a major problem for doctors and patients, even when the transplants are human to human. There are precious few examples of any long-term organ xenotransplants succeeding.

There has never been a greater need for new transplantation options. In the United States alone, more than 124,000 patients are currently on the waiting list for transplants, yet only approximately 28,000 procedures are carried out annually. It's been estimated that a new individual is added to the national transplant list every ten minutes and that an average of twenty-two people a day die while waiting for a transplant or become so sick that they are no longer eligible to receive a transplant. The shortage of donor organs is the biggest cause of this ongoing tragedy.

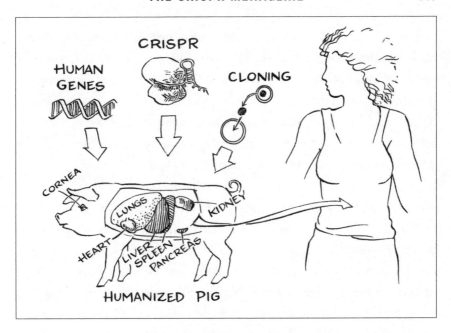

Xenotransplantation using humanized pigs

New technologies including CRISPR provide a way to generate pigs with organs suitable for human transplant. Previous advances focused on transferring human genes into the pig genome so that pig organs might escape the hyperacute immune rejection that threatens any xenotransplant. Gene editing is now being harnessed to shut down pig genes that might provoke the human immune response and to eliminate the risk that porcine viruses embedded in the pig genome could hop over and infect humans during transplantation. Finally, cloning technologies offer a way to seamlessly combine the different genetic alterations into a single animal. As the CEO of one prominent company in this field has stated, the goal is to provide "an unlimited supply of transplantable organs," organs that can be produced on a made-to-order basis.

It's still early days for this effort, but records have already been shattered using pigs that were humanized through genetic engineering: a transplanted kidney lasted over six months in a baboon recipient, and a pig heart was similarly accepted by a baboon's body for two and a half years. Tens of millions of dollars have been committed to future research, and a company called Revivicor has already outlined plans to breed a thousand pigs a year in state-of-the-art facilities with surgical theaters and helipads to deliver fresh organs whenever they're needed. It seems only a matter of time before xenotransplantation makes its way into clinical trials — and until CRISPR opens a new door for patients in dire need of new organs and new drugs.

As someone who grew up surrounded by the plants and animals shaped by Hawaii's vibrant ecosystem, I'm both fascinated and, admittedly, a bit apprehensive about all the ways CRISPR is being used to genetically modify animals. I hope that gene-edited livestock will make agriculture more humane and environmentally friendly, not just more profitable. Gene-edited animal models like mice and monkeys will advance our understanding of human disease, and gene-edited pigs may serve as future organ donors, but I hope that a shared respect for animal welfare will temper these and similar efforts.

But with gene editing enabled by CRISPR, it seems inevitable that some people will edit animal genes without any medical purpose and without the goal of making animal agriculture any more sustainable, productive, or humane. Take the case of a brand-new breed of miniaturized pig — the so-called micropig. Created with gene editing at the Beijing Genomics Institute (BGI) in China, these adorable swine wowed crowds at the biotech summit where they were first unveiled. The adult micropigs weigh in at about thirty pounds, similar to a medium-size dog, whereas farm pigs often weigh more than two hundred pounds. BGI originally created the micropigs for research purposes, since the large size of normal pigs can make them burdensome for lab workers. By slicing apart and inactivating the gene that responds to growth hormone, scientists produced stunted pigs that otherwise developed normally. And although the micropigs remain helpful for research — a Chinese group recently used CRISPR in micropigs to generate a human Parkinson's disease model — BGI has also begun offering them as pets, with a price tag of around fifteen hundred dollars per pig. Someday, consumers might even have the option of selecting custom features, like varied colors or coat patterns, all made possible with gene editing.

Some bioethicists, like Jeantine Lunshof at Harvard Medical School, are concerned about genetic manipulation being used "for the sole purpose of satisfying idiosyncratic aesthetic preferences of humans." Yet I'm not convinced this is categorically a bad thing. After all, at any ordinary dog park, you could find a four-pound Chihuahua frolicking next to a two-hundred-pound Great Dane — both members of the same species. What is breeding but another tool of genetic manipulation, like CRISPR, only less predictable and efficient? There's even a case to be made that this kind of genetic manipulation is better than breeding. Unlike micropigs, whose health is no different than their normal-size relatives, extensive inbreeding of dogs has had devastating health consequences. Labradors are prone to some thirty genetic conditions, 60 percent of golden retrievers succumb to cancer, beagles are commonly afflicted with epilepsy, and Cavalier King Charles spaniels suffer from seizures and persistent pain due to their deformed skulls. These poignant medical problems haven't kept humans from letting tastes dictate the genotype and phenotype of humankind's best friend.

Whatever one thinks of them, gene-edited cats and dogs — created with the help of biotechnology — are just around the corner. In late 2015, scientists in Guangzhou, China, reported the first application of CRISPR in beagles, using it to enhance muscle mass by knocking out the same *myostatin* gene linked to double muscling in whippet dogs and Belgian Blue cows. The two puppies that contained the intended mutations were named Hercules and Tiangou, in honor of the superhuman hero of Greek mythology and the heavenly dog in Chinese mythology. Although one of the lead scientists asserted that the extra-muscular beagles weren't going to be bred as pets and would instead facilitate the use of dogs for biomedical research, he noted the potential advantages of extra muscle for police and military applications. The team concluded their study by noting that CRISPR could "also promote the creation of new strains of dogs with favorable traits for other purposes."

With easy-to-use gene editing, it surely won't be long before consumers can order off-the-shelf enhancements to any dog breed. Where else will our imaginations take us? If genetic manipulation succeeded in dehorning cows, why not use it to intentionally horn horses? And if we're thinking about adding appendages, why stop there? Researchers at UC Berkeley used CRISPR to generate a bizarre array of bodily transformations in crustaceans — gills growing where they shouldn't be, claws becoming legs, jaws turning into antennae, and swimming limbs becoming walking limbs. Scholars and journalists have already started dreaming that CRISPR might be used to create mythical creatures like winged dragons by editing the genes of Komodo dragons, noting in a prominent bioethics journal that, while basic physics would prevent them from breathing fire, "a very large reptile that looks at least somewhat like the European or Asian dragon (perhaps even with flappable if not flyable wings) could be someone's target of opportunity."

While some scientists may harness CRISPR to create mutated creatures that never before existed, others are applying CRISPR to resurrect native creatures that no longer exist, a pursuit aptly called de-extinction. It predates CRISPR by a few decades, and gene editing is just one of the methods that scientists hope will make it possible. In cases where an extinct species' traits are shared by contemporary descendants, scientists might be able to turn the latter into the former via selective breeding, creating an animal reminiscent of the extinct creature. This strategy is being undertaken in Europe to bring back the auroch, a wild ox that went extinct in the early 1600s, and in the Galápagos Islands to resurrect a species of saddleback tortoise from Pinta Island whose last known member died in 2012. In cases where tissues from extinct animals have been carefully preserved, cloning is another possibility. For instance, the

Pyrenean ibex, a wild goat, died out in 1999, but cryogenic preservation of skin biopsies taken from the last living specimen allowed Spanish scientists to implant its genetic material into the egg of a domestic goat. (The same procedure was used to clone Dolly the sheep in 1996.) With the live birth that resulted, the scientists achieved the first-ever resurrection of an extinct animal, though, regrettably, the newborn died just minutes after birth. The same cloning approach is now being pursued by Russian and South Korean scientists who are hoping to use mammoth tissues recovered in eastern Russia to resurrect woolly mammoths.

CRISPR offers another way to bring bygone species back to life — one not so different from the fictional depiction of dinosaur de-extinction in the book and subsequent 1993 Hollywood film *Jurassic Park*. In that compelling science fiction tale, scientists spliced into frog DNA the genes of deceased dinosaurs that they'd recovered from fossilized mosquitoes preserved in amber. Sadly (or luckily, depending on how you feel about dinosaurs), the chemical bonds in DNA are far too unstable to remain intact for sixty-five million years. But author Michael Crichton wasn't far off the mark with this idea.

A similar strategy is being pursued for woolly mammoths by a team of Harvard researchers led by George Church. A key starting point is the high-quality, fully sequenced genome that was obtained from two woolly mammoth specimens that died some twenty to sixty thousand years ago; the genomes have allowed scientists to exhaustively analyze the precise DNA changes between the mammoth and the modern-day elephant, its closest relative. Not surprisingly, given woolly mammoths' icy habitats, the 1,668 genes that differ between the two genomes encode proteins whose functions relate to temperature sensation, skin and hair development, and production of fat tissue. In 2015, working with elephant cells, Church's team used CRISPR to convert the elephant variant to the woolly mammoth variant for fourteen of those genes, and sequential gene editing could theoretically accomplish the same for the remainder.

Fully morphing the elephant genome into the woolly mammoth genome would involve changing over 1.5 million DNA letter differences between them, and there are no guarantees that edited elephant cells could be used to establish an actual pregnancy. Even if it could be done, would the resulting animal, birthed by an elephant and lacking its original environment and social culture, really be a woolly mammoth? Or would it simply be an elephant with new traits inspired by woolly mammoth genetics?

Ever since I first heard about experiments like these, I've struggled to decide whether they're admirable, deplorable, or something in between. In my mind, as in the minds of most people in the broader scientific community, the jury is still out. One thing seems clear: some of the uses to which CRISPR has been put in the animal kingdom are more noble than others, and each time I set out to determine how I feel about a particular one, I find myself plunging into a thicket of arguments and counterarguments.

What, really, is the point of resurrecting the woolly mammoth, or any other extinct species, for that matter? One reason may well be wonder —a feeling of surprise and admiration at the possibilities afforded by nature and science performed at the most advanced level. Some people flock to zoos or go on safaris to watch lions and giraffes up close; imagine what a fascinating, even emotional experience it would be to stand face to face with a real-life mammoth. Other motivations for editing the elephant genome to be more woolly mammoth–like include saving the endangered Asian elephant species and reducing carbon release from tundra.

There is also an ethical argument for de-extinction. If we've driven a species to extinction and we now have the power to bring it back, do we have a duty to do so? One of the organizations leading the de-extinction movement, the Long Now Foundation, thinks so; its mission is to "enhance biodiversity through the genetic rescue of endangered and extinct species" using the tools of genetic engineering and conservation biology, and it engages in both de-extinction and extinction-prevention efforts. On its list of candidates for de-extinction are passenger pigeons, which were eradicated by hunting in the nineteenth century; great auks, whose populations plummeted in the sixteenth century because humans slaughtered them for their down; and gastric-brooding frogs, which were extinguished around 1980 by pathogenic fungi introduced into their native habitat by humans.

Yet it is far from certain that de-extinct species would be received hospitably by the modern world, or that reintroducing them would be free of risk — for them or for us. In the same way that living species released into foreign environments can wreak ecological havoc on their new habitats, de-extinct species could badly disrupt ecosystems they're released into. Because we've never been able to bring back an extinct species, moreover, there's no telling how big the shock waves from their reappearance might be or where they might end.

There are other good reasons to oppose the use of CRISPR to resurrect extinct species, reasons similar to the arguments against using it to create designer pets: we have to consider morality and animal welfare. How much animal suffering — such as the deformities and premature deaths that commonly accompany cloning procedures — can we justify in the pursuit of scientific research that almost certainly will never influence or improve human health? Will focusing on de-extinction and designer pets distract us from protecting existing endangered species or mistreated and neglected companion animals? And on a more basic level, if we can avoid altering nature more than we already have, shouldn't we try to do so?

CRISPR is forcing us to confront difficult, perhaps unanswerable questions like these, many of which boil down to conundrums about the relationship between humans and nature. Humans have been changing the genetic makeup of plants and animals since long before the advent of genetic engineering. Should we refrain from influencing our environment with this new tool even though we haven't showed such restraint in the past? Compared to what we've done to our planet already, whether intentional or not, is CRISPR-based gene editing any less natural or any more harmful? There are no easy answers to these questions.

There is one way, at least, in which the power to edit the genes of other species could prove to be more dangerous than any changes humans have made to the planet so far. I'm referring to a revolutionary technology known as a gene drive, so called because it gives bioengineers a way to "drive" new genes — along with their associated traits — into wild populations at unprecedented speeds, a kind of unstoppable, cascading chain reaction.

With gene drives, as with other developments in the burgeoning gene-editing field, the science has moved so fast that it's hard to keep up. Just a year after it was first proposed in a theory paper, CRISPR gene drives proved effective, first in fruit flies and then in mosquitoes. These gene drives harness the power of a special type of inheritance pattern, one that defies the normal way genetic information is shared between generations of living things.

In normal sexual reproduction between species that contain two copies of each chromosome, offspring acquire just one chromosome copy from each parent, meaning that any particular gene variant has a 50 percent probability of being inherited. However, there are certain DNA sequences, called selfish genes, that can increase their frequency in the genome with each generation, even without conferring any fitness advantage on the offspring. In 2003, evolutionary biologist Austin Burt proposed a way to harness selfish genes in order to spread novel traits more efficiently and to ensure that offspring would have a 100 percent probability of inheriting a given segment of DNA. But his idea hinged on a technology that didn't really exist at the time: easily programmable DNA-cutting enzymes that would allow for simple gene editing.

Enter CRISPR. In the summer of 2014, George Church's team at Harvard, led by Kevin Esvelt, proposed a way to design and build gene drives with the help of efficient gene editing. In essence, the idea relies on a gene knock-in approach, in which scientists use CRISPR to cut DNA at an exact location and insert a new sequence of letters into the breach. There is one major difference with a gene drive, however: part of the new DNA added in contains the genetic information that encodes CRISPR itself. Like that sci-fi trope of a self-replicating machine, a CRISPR gene drive can autonomously copy itself into new chromosomes, allowing it to grow exponentially within a population. By combining CRISPR with various genetic payloads, such as pathogen-resistance genes, Esvelt theorized, scientists could program CRISPR to copy not only itself, but any other desirable DNA sequences.

As it turns out, gene drives can be as remarkably effective as the theory predicts. In early 2015, Ethan Bier and his student Valentino Gantz at UC San Diego reported the first successful demonstration of a CRISPR gene drive in the common fruit fly, using it to drive a defective pigmentation gene into the genome. The result: 97 percent of the edited flies were a new, light yellow color instead of the species' usual yellow-brown. Within half a year, the same team had extended their initial proof-of-concept results in fruit flies to promising tests with mosquitoes. Rather than simply changing the bugs' color, however, this new gene drive spread a gene that gave the offspring resistance to *Plasmodium falciparum*, the parasite responsible for hundreds of millions of malaria infections annually. The rate of success in the wild mosquitoes on which this new gene drive was tested was even higher: 99.5 percent.

If the first of these gene drives (for pigmentation) seems benign and the second (for malaria resistance) seems beneficial, consider a third example. Working independently of the California scientists, a British team of researchers — among them Austin Burt, the biologist who pioneered the gene drive concept — created highly transmissive CRISPR gene drives that spread genes for female sterility. Since the sterility trait was recessive, the genes would rapidly spread through the population, increasing in frequency until enough females acquired two copies, at

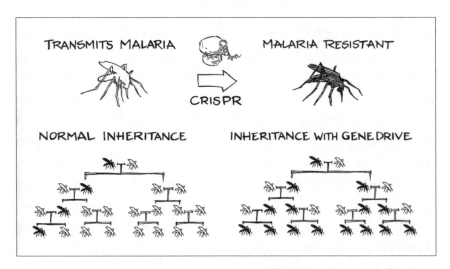

Using CRISPR to build gene-drive mosquitoes

which point the population would suddenly crash. Instead of eradicating malaria by genetically altering mosquitoes to prevent them from carrying the disease, this strategy presented a blunter instrument — one that would cull entire populations by hindering reproduction. If sustained in wild-mosquito populations, it could eventually lead to outright extermination of an entire mosquito species.

It's not the first time that scientists have turned to genetic engineering to reduce insect populations. A common practice used for decades involves the release of sterilized males into the environment; the technique has all but eliminated certain agricultural pests through North and Central America. Another approach being developed by a British company called Oxitec involves inserting a lethal gene into the mosquito genome, and field trials have already commenced in Malaysia, Brazil, and Panama. However, these strategies are inherently self-limiting; the genetic alterations are rapidly eliminated by natural selection, and the only way to make a dent in mosquito populations is to repeatedly release large batches of the modified insects.

CRISPR gene drives, by contrast, are self-sustaining; since the mode of inheritance appears to outsmart natural selection, the modified insects propagate and pass on their defective traits indefinitely. This thoroughness is what makes gene drives so powerful — and so alarming. It's been estimated that, had a fruit fly escaped the San Diego lab during the first gene drive experiments, it would have spread genes encoding CRISPR, along with the yellow-body trait, to between 20 and 50 percent of all fruit flies worldwide.

Scientists pursuing CRISPR gene drives have been outspoken about the need to carefully weigh the risks before conducting further experiments and about the importance of developing guidelines that ensure future research proceeds safely. Perhaps the most obvious safeguard to prevent accidentally unleashing a gene drive into the world is stringent containment, such as physical barriers separating organisms from the environment and ecological barriers between the animal's habitable range and the geographical location of the laboratory. At a recent conference at which Ethan Bier presented his research, he showed the audience pictures of the extensive containment procedures in place to prevent the accidental release of test insects. But if all else fails, scientists have proposed a variety of strategies that could theoretically inactivate gene drives that run amok. One of these is the so-called reversal drive, a gene drive that essentially functions as an antidote by overwriting any changes in the genome created by the original gene drive.

Even with the most cautious experimental design and planning, we can't predict all of the environmental effects that a gene drive might have, and we can't completely eliminate the possibility of a gene drive getting out of control and disrupting an ecosystem's delicate balance. These risks were reflected in a recent report authored by the National Academies of Sciences, Engineering, and Medicine, which endorsed ongoing research and limited field trials but stopped short of recommending that gene drives be released into the environment.

There's also no way to guarantee that this incredibly powerful tool won't wind up in the hands of people who have no compunction about using gene drives to cause harm — and who may, indeed, be attracted to them for exactly that purpose. The ETC Group, a biotech watchdog organization, worries that gene drives — what they call "gene bombs" — could be militarized and weaponized to target the human microbiome or major food sources.

But as frightening as gene drives could be, we might find it impossible to justify keeping them locked away in the lab. As Austin Burt wrote, "Clearly, the technology described here is not to be used lightly. Given the suffering caused by some species, neither is it obviously one to be ignored." Gene drives could help us address global problems in agriculture, conservation, and human health in a far more targeted way than previous approaches have allowed. Among the applications that have been proposed are reversing the genetic causes of herbicide and pesticide resistance that have evolved among organisms that threaten agriculture; promoting biodiversity by controlling, even eradicating, invasive species populations like Asian carp, cane toads, and mice; and stamping out infectious diseases such as Lyme disease, which is caused by certain bacteria transmitted by ticks, and schistosomiasis, caused by flatworm parasites transmitted by aquatic snails. But the most momentum, by far, is in the push to use gene drives to target the mosquito.

The mosquito causes more human suffering than any other creature on earth. Mosquito-borne diseases — malaria, dengue virus, West Nile virus, yellow fever virus, Chikungunya virus, Zika virus, and many others — have an annual death toll in excess of one million. CRISPR-based gene drives might be the best weapon we have against this pervasive threat, whether we use them to prevent mosquitoes from harboring specific pathogens or to wipe out the insects altogether. On top of that, genetic strategies like CRISPR might be safer than toxic pesticides, and they offer the allure of solving biological problems with biology.

Would it be a blessing or a curse to suddenly be rid of the winged pests that have inhabited the earth for more than one hundred million years?

Somewhat incredibly, scientists don't seem overly concerned about a world without mosquitoes. As one entomologist put it, "If we eradicated them tomorrow, the ecosystems where they are active will hiccup and then get on with life." If he's right and we could have a world free from the ravages of mosquito-borne illness, can we justify *not* taking the risk?

I pose these questions because I, too, am searching for answers. The stakes are high enough to make these some of the most pressing scientific issues facing us today. It is vital that we all weigh in on how this new biotechnology should be used in the plant and animal worlds. With concerted education and soul-searching, I'm hopeful we can answer these questions — that we can benefit from gene-edited flora and fauna while avoiding the biggest pitfalls.

Still, like many scientists, I sometimes can't help but view the work being done with plants and animals as a sort of dry run for the ultimate goal of gene editing. I'm referring, of course, to the idea that Emmanuelle and I had when we first contemplated the outcome of our research collaboration: the dream that, someday, our work would help rewrite the DNA in human patients to cure disease.

6

TO HEAL THE SICK

AS 2015 WOUND DOWN, I was completing the usual end-of-semester chores: assigning grades to my students and generating project budgets and research goals for the year ahead. At the same time, however, I was also preparing for a very different kind of task: a presentation I would soon give with Vice President Joe Biden at the annual meeting of the World Economic Forum in Davos, Switzerland, in January 2016.

The invitation to speak alongside the vice president had been the latest vote of confidence in CRISPR's potential as a medical tool. I had already been planning to travel to Davos, where civic and private-sector leaders assemble each winter to discuss issues of pressing global importance. It would be my second time attending the meeting, and on this visit, like my previous one, I had been asked to speak about CRISPR technology and its global economic and social impacts, including the effects it would have on the world of medicine.

But Vice President Biden's invitation was perhaps the most resounding affirmation of this technology's significance for the field of public health. Just as powerful as the implications for CRISPR research was the reason behind the invitation: Biden would be holding a press conference where he, along with scientists and clinicians, would unveil an initiative by President Obama to coordinate efforts to treat and cure cancer. In the tradition of the 1960s American space program that resolved to — and, in short order, did — send humans to the moon, this "cancer moonshot" aimed to rally the country's best and brightest minds to find cures for cancer in all its guises. The fact that Biden's son Beau had recently passed away after a years-long battle with brain cancer only made the occasion more compelling and brought home the human tragedy and indiscriminate pain that cancer causes so many families.

Although it took some wangling, I was able to recruit a colleague to cover my January teaching commitments and travel to Davos early to participate in the Biden meeting, which turned out to be as fascinating as it was affecting. I learned a great deal from my fellow attendees, many of them scientists engaged in cancer-related research, drug development, and clinical practice. As they shared the latest findings from this outsize corner of the medical world, their revelations made clear to me how far we'd come in treating cancer since my father's all-too-brief struggle with melanoma in 1995. It also drove home the distance we still had to go to find effective cancer treatments, let alone cures, and reminded me yet again of how CRISPR could accelerate that process.

At the press conference, as I discussed the CRISPR technology and the implications of this tool for cancer treatment, I gazed out at the bank of TV cameras and crowd of journalists in attendance. Suddenly I felt like I was watching the event from the vantage point of the reporters, wondering what an RNA biochemist was doing sitting next to doctors who had dedicated their careers to finding cancer cures. I was both honored to be there and humbled by the thought of just how far I had traveled, both literally and metaphorically, to be discussing such a consequential public-health issue alongside the vice president of the United States.

Among politicians and scientists, and increasingly among the public as well, there is growing appreciation of the important role gene editing might play in developing new treatments or even cures for disease. In addition to federal support for such therapies, in the form of grants to academic researchers, the private sector is getting involved. Three startup therapeutics companies, two based in Cambridge, Massachusetts, and one in Basel, Switzerland, have been co-founded by academic scientists, including Emmanuelle and me, and aided by hundreds of millions of dollars in venture capital backing. As of this writing, all three have already become publicly traded companies. The University of Pennsylvania is conducting the first CRISPR-based clinical trial to be approved in the United States, with financial backing from Internet billionaire Sean Parker. A new biotech institute in San Francisco, affiliated with UC Berkeley, UC San Francisco, and Stanford University, is benefiting from a generous contribution of more than half a billion dollars from Facebook's Mark Zuckerberg and his wife, pediatrician Priscilla Chan. And in the Bay Area, I had the privilege of launching the Innovative Genomics Institute, aimed at harnessing technologies like CRISPR to lead the revolution in genetic engineering and fight against disease.

If these recent examples are any indication, the future of medicine will increasingly feature CRISPR and increasingly reflect new alliances and partnerships between public and private sponsors. But we won't need to wait to see the power of CRISPR in preventing disease. The evidence is right in front of us.

Preclinical work in animal models is already demonstrating CRISPR's incredible ability to hunt down and repair mutated genes inside living creatures. In December of 2013, less than a year after several labs, including my own, reported the successful use of bacteria-derived CRISPR molecules in human cells for gene editing, a team of Chinese researchers programmed the same CRISPR molecules to find and fix a single-letter mutation among the 2.8 billion DNA letters of the mouse genome. In so doing, they performed the first outright, CRISPR-based cure of a genetic disease in a live animal.

I was electrified by this news, although I can't say that I was surprised, given how rapidly the technology was being implemented. Still, the researchers' accomplishment represented something momentous: It was the first of a new breed of exquisitely precise genetic therapies and seemed to mark the beginning of a new era in medicine — one in which at least some of the more than seven thousand human genetic diseases caused by a defined, single-gene mutation might be cured, thanks to a one-size-fits-all molecular tool.

The proof-of-principle experiments from China had cured a mouse of congenital cataracts, a disorder in which a defective gene causes eye cloudiness and a decline in vision. Over the next couple of years, scientists used CRISPR to cure live mice of muscular dystrophy (a severe muscle-wasting disease), as well as various metabolic disorders affecting the liver. Meanwhile, working in cultured human cells that were often derived from patient tissue samples, hundreds of researchers used CRISPR to repair an ever-expanding number of DNA mutations associated with some of the most devastating genetic diseases out there, everything from sickle cell disease and hemophilia to cystic fibrosis and severe combined immunodeficiency. Whether the underlying problem was incorrect letters of DNA, missing letters, extraneous letters, or even large chromosomal abnormalities, it seemed that no single-gene error was too great for CRISPR to fix.

The potential utility of therapeutic gene editing goes far beyond simply reverting mutated genes back to their healthy states. Some scientists are employing CRISPR in human cells to block viral infections, just like this molecular defense system naturally evolved to do in bacteria. In fact, the first clinical trials to use gene editing are aimed at curing HIV/AIDS by editing a patient's own immune cells so the virus can't penetrate them. And in another landmark effort, the first human life was saved by gene editing in combination with another emerging breakthrough in medicine: cancer immunotherapy, in which the body's own immune system is trained to hunt down and kill cancerous cells.

It's easy to get caught up in the excitement. The fact that gene editing might be able to reverse the course of a disease — permanently — by targeting its underlying genetic cause is thrilling enough. But even more so is the fact that CRISPR can be retooled to target new sequences of DNA and, hence, new diseases. Given CRISPR's tremendous potential, I've grown accustomed over the past several years to being approached by established pharmaceutical companies asking for my help in learning about the CRISPR technology and about how it might be deployed in the quest for new therapeutics.

But therapeutic gene editing is still in its infancy — indeed, clinical trials have only just begun — and there are still big questions about how things will progress from here. The decades-long struggle to make good on the promise of gene therapy should serve as a reminder that medical advances are almost always more complicated than they might seem. For CRISPR, too, the road leading from the lab to the clinic will be long and bumpy.

Deciding what types of cells to target is one of the many dilemmas confronting researchers — should they edit somatic cells (from the Greek *soma*, for "body") or germ cells (from the Latin *germen*, for "bud" or "sprout")? The distinction between these two classes of cells cuts to the heart of one of the most heated and vital debates in the world of medicine today.

Germ cells are any cells whose genome can be inherited by subsequent generations, and thus they make up the germline of the organism — the stream of genetic material that is passed from one generation to the next. While eggs and sperm are the most obvious germ cells in humans, the germline also encompasses the progenitors of these mature sex cells as well as stem cells from the very early stages of the developing human embryo.

Somatic cells are virtually all the other cells in an organism: heart, muscle, brain, skin, liver — any cell whose DNA cannot be transmitted to offspring.

Mouse geneticists (and animal breeders in general) have jumped at the chance to alter germ cells using CRISPR. That's because editing the germline is the easiest way to demonstrate the technology's curative power. Normally, by the time a mouse with a disease-causing genetic mutation reaches adulthood, it's too late to correct the error; what began

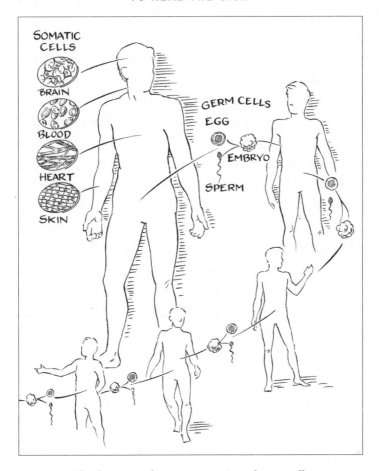

The distinction between somatic and germ cells

as a mistake in a single fertilized egg cell has been copied into billions of descendant cells, making it all but impossible to stamp out every last trace of the disease. (Imagine trying to correct an error in a news article after the newspapers have been printed and delivered, as opposed to when the article is still just a text file on the editor's computer.) By focusing on the germline, scientists can send CRISPR into the embryo at its earliest stage of development and reverse the mutation in a single cell. As the embryo develops into an adult organism, the repaired DNA is faithfully copied into every daughter cell, including the germ cells that will eventually transmit the genome to subsequent generations.

But while germline editing has been useful as a research tool in lab mice, its use in humans poses significant safety and ethical challenges. Should we really be manipulating the genome of unborn individuals and tweaking the *Homo sapiens* gene pool in a way that cannot be easily reversed? Are we as a species prepared to assume control of — and responsibility for — our own evolution, purposefully mutating our genomes rather than leaving their makeup to chance? These are enormous, thorny issues, and I'll turn to them fully in the final two chapters of this book.

Ethically speaking, editing somatic cells to treat genetic disease is much more straightforward than editing germ cells, since the changes can't be passed down to the patient's descendants. Practically, however, it's much more complex. Reversing a disease-causing mutation in a single human germ cell is much simpler than trying to do the same thing inside some of the fifty trillion somatic cells that make up a human body. To pull that off, scientists have to solve a host of new problems — but solve them we must if we're to help the many men, women, and children afflicted with genetic diseases. In those cases, editing germ cells will do nothing to ease their suffering. It's far too late for that. Somatic editing is the only way.

It might be hard to imagine that gene editing can reverse the course of a disease in any human, much less an adult who's been living with it for his or her entire life. The roots of the disease run deep by that point, and changing a patient's DNA might not undo the accumulated effects of a faulty genetic code.

To be sure, there are limits to what we'll be able to do with CRISPR in this regard. Some diseases don't have clear-cut genetic causes, and in some conditions, like schizophrenia and obesity, genetics play a complex role, with many genes implicated but each one contributing only a small effect. Given how hard it will be to use CRISPR to edit just a single gene in the human body safely and effectively, we're not likely to begin editing multiple genes at once any time soon.

CRISPR offers the greatest hope to treat monogenic genetic diseases — those caused by a single mutated gene. At a basic level, these diseases result when the mutated gene produces either a defective protein or no protein at all. If gene editing succeeds in restoring normal production of the healthy protein before the mutation causes any irreversible damage, it should serve as a one-time intervention whose therapeutic effects endure for the rest of the patient's life. This stands in contrast to existing treatments for genetic disease that often rely on temporary solutions involving transplants or repeated drug administration. Importantly, physicians won't need to edit all the cells in a patient's body in order to heal a genetic disease. Even though all cells possess the disease-causing DNA mutation, the symptoms often manifest themselves only in those tissues where normal functioning of the mutated gene matters most. For example, immunodeficiencies mostly affect white blood cells; Huntington's disease primarily affects neurons in the brain; sickle cell disease affects only hemoglobin-carrying red blood cells; and cystic fibrosis does most of its damage in the lungs. Since the effects of genetic diseases tend to be localized in this way, therapies will need to treat cells in the most affected parts of the body.

That's not to say it'll be easy to get CRISPR to these locations, much less get it inside the cells themselves. This delivery problem is one of the greatest challenges that somatic gene-editing technologies will have to face.

The available delivery strategies can be broken down into two major categories: in vivo gene editing (from the Latin for "within the living," as mentioned earlier) and ex vivo gene editing (from the Latin for "out of the living"). In the former approach, CRISPR is sent directly into the body of the patient to do its work onsite; in the latter, the patient's cells are edited outside of the body and then placed back into the patient. Ex vivo therapy is the far simpler approach, and since scientists have already

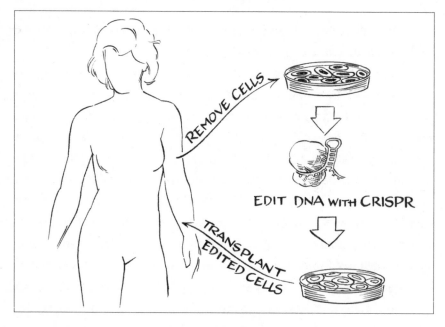

Ex vivo CRISPR therapy

mastered editing cells in the lab, we're one step closer to this mode of therapy than to the in vivo kind. Another benefit of ex vivo gene editing is that the gene-edited cells can be subjected to rigorous quality-control testing before ever seeing a patient's body.

Because ex vivo gene editing requires doctors to first remove diseased cells from the body, it is uniquely suited to treating blood-based diseases. Using a combination of gene editing, blood donation, and transfusion techniques, doctors can take the affected blood cells from the patient's body, edit them using CRISPR, and return them to the circulatory system.

Two of the promising disease targets for ex vivo CRISPR therapies are sickle cell disease and beta-thalassemia. Among the most common genetic diseases, they both result from molecular defects in hemoglobin, the major protein component of red blood cells and the one that transports oxygen from the lungs to the tissues of the body. The sources of these molecular defects are DNA mutations in the beta-globin gene, which encodes one of the two unique protein chains that make up the hemoglobin molecule.

Sickle cell disease and beta-thalassemia can actually both be cured by bone marrow transplantation. When physicians transplant the bone marrow from a healthy individual into a sick patient, the abundant blood stem cells in the marrow produce healthy new red blood cells for the rest of the patient's life. The problem with this sort of stem cell transplantation, however, is that there aren't nearly enough donors who both match the recipient immunologically and are willing to undergo the invasive procedure. Even when a matching donor is found and the patient's body accepts the transplanted cells, the procedure is still risky; many patients develop graft-versus-host disease, a sort of reverse immunological reaction, which can be fatal.

Gene editing may solve this problem by allowing patients to serve as both recipient *and* donor of the stem cells. If doctors can isolate stem cells from a patient's bone marrow, repair the cells' mutated beta-globin genes with CRISPR, and then return those edited cells to the patient, they won't have to worry about donor availability or the risk of an immunological clash between the patient's body and the transplanted cells. Numerous laboratories have already convincingly demonstrated that patient cells can be precisely repaired in the laboratory and that those edited cells churn out robust amounts of healthy hemoglobin; researchers have even shown that the edited human cells can function inside immunocompromised mice. Multiple academic research teams as well as commercial companies are now working to make the procedure available to human patients.

There's good reason to be optimistic about such clinical trials with ex vivo gene editing, given recent developments in the related field of ex vivo gene therapy. (Remember that gene editing repairs mutated genes directly in the genome, whereas gene therapy splices new, healthy genes into the genome.) The biotechnology company Bluebird Bio has been developing a product that treats beta-thalassemia and sickle cell disease by inserting new beta-globin genes into blood stem cells, and GlaxoSmithKline has similarly built an effective gene therapy drug that cures severe combined immunodeficiency (SCID) by inserting the missing gene into the genome. In both approaches, the general intervention strategy is the same: remove the patient's cells, correct them in a test tube, then put them back into the patient. Gene editing is likely to be the safer approach, though, since it perturbs the genome as little as possible.

The first-ever clinical trial to demonstrate ex vivo gene editing in human subjects showed just how promising, and powerful, this procedure can be. Ironically, the target wasn't a genetic disease at all but the human immunodeficiency virus (HIV). And although this clinical trial was developed before the CRISPR technology came into existence — it utilized the zinc finger nuclease (ZFN) technology described in chapter 1 — its success bodes well for the prospect of using gene editing to combat this pandemic and also to treat many genetic diseases.

Believe it or not, some lucky people are naturally resistant to HIV. These individuals lack thirty-two letters of DNA in the gene for a protein called CCR5, which is located on the surface of white blood cells — those cells that form the bedrock of the body's immune system. CCR5 proteins are one of the parts of the cell's surface that the HIV virus latches onto in the initial stage of its invasion. This specific, thirty-two-letter deletion causes the CCR5 protein to be truncated and prevents it from making its way to the cell surface. Without CCR5 proteins to attach to, HIV molecules can't infect the cells.

In people of African and Asian descent, the thirty-two-letter *CCR5* deletion is virtually nonexistent, but it's fairly prevalent among Caucasian people; 10 to 20 percent of Caucasians possess one copy of the mutated gene, and homozygous individuals — those who possess two copies — are completely resistant to HIV. Roughly 1 to 2 percent of Caucasians worldwide (most of them in northeastern Europe) are fortunate enough to have this trait. These CCR5-lacking individuals are otherwise com-

pletely healthy and even experience a reduced risk of certain inflammatory diseases; the missing protein doesn't cause any adverse effects. Just
about the only known risk to not having it is a possible increase in susceptibility to the West Nile virus.

Unsurprisingly, the pharmaceutical industry has devoted vast resources to developing drugs that disrupt the interaction between HIV
and CCR5 in the hope of protecting people who aren't lucky enough to
have two copies of the thirty-two-letter deletion in their genomes. But
recent studies have demonstrated conclusively that we can accomplish
the same thing—that is, prevent HIV from latching onto CCR5—by
editing the *CCR5* gene itself. Multiple labs have already pulled this off
using CRISPR, at least with cells in a petri dish. But credit for the first
success in editing the *CCR5* gene in human subjects goes to the ZFN
technology and to a California-based company called Sangamo Therapeutics.

Working with physicians at the University of Pennsylvania, Sangamo
researchers conducted a clinical trial using a gene-editing drug that
knocked out the *CCR5* gene. The early-phase trial was aimed primarily
at testing the drug's safety; the researchers wanted to know if the edited
cells, whose DNA had been modified in the lab, would be accepted by
the patients' bodies without major side effects. As it turned out, the study
would also show just how effective gene editing could be in reversing the
course of a disease.

All of the twelve HIV-positive patients who enrolled in the Sangamo
study first had a sample of their white blood cells extracted from their
blood. These white blood cells were cleaned up in the lab and edited
with a ZFN that targeted and cut the 155th letter in the *CCR5* gene. Because the sliced-up gene was repaired using error-prone end joining, the
resulting changes were sufficient to inactivate the gene and prevent any
functional CCR5 protein from being produced. Next, the edited cells
were allowed to multiply in the lab. Finally, each patient was reinfused

with his or her own edited cells and then monitored over the course of roughly nine months.

The researchers who conducted the trial concluded that infusions of *CCR5*-modified immune cells "are safe within the limits of this study." Not a jaw-dropping finding, perhaps, but nevertheless an encouraging sign that gene editing could be used therapeutically in patients — at least ex vivo, by way of cells cultured and treated in the laboratory. And buried in the results section of the study were even more promising data. The physicians found that, in addition to showing long-lived persistence in the body (a sign that the transplanted cells had readily engrafted and proliferated), the edited cells caused HIV levels to rebound more slowly than normal when antiretroviral therapy was temporarily interrupted. In other words, there were clear signs that the ZFN treatment had succeeded in reducing the infection, not like a typical drug, but with single-letter changes to the patients' genomes.

Although ZFNs have had a head start, CRISPR has already been used to explore several possible therapies aimed at eliminating HIV. One approach involves programming CRISPR to target genetic material from the HIV virus, ridding patients' cells of HIV by literally snipping the infectious DNA out of their genomes. Yet another method might be best described as "shock and kill": it uses a deactivated form of CRISPR to intentionally awaken the dormant virus so that it can be targeted using existing drugs.

It's become clear that the clinical possibilities for ex vivo gene editing are enormous, whether harnessed to treat genetic diseases or viral infections. But of course, not all illnesses are rooted in the blood. For diseases that afflict the body's tissues, physicians can't rely on therapies that require the removal and reinstatement of affected cells; the procedures for doing so are simply too invasive and too risky. To treat these diseases, we'll need to deliver CRISPR into the patient's body, to the tissue where the disease is exerting its greatest effect. And while we still have a long way to go before it will be possible to offer this therapy to patients, the advances being made in this area constitute some of the most incredible developments in medical science I've ever seen.

Before doctors can treat patients with in vivo gene-editing therapies, scientists will need to solve numerous problems that ex vivo approaches neatly circumvent. Physicians must figure out how to get CRISPR into the tissues that are most affected by a given disease. In addition, this must be accomplished without provoking an immune response in the patients' bodies. Furthermore, Cas9 and its RNA guide will have to be stable enough to survive inside the body until editing is completed.

To address these challenges, some CRISPR researchers are turning to one of their favorite delivery vehicles: viruses. Viruses are incredibly adept at sneaking genetic material inside host cells; after all, they've

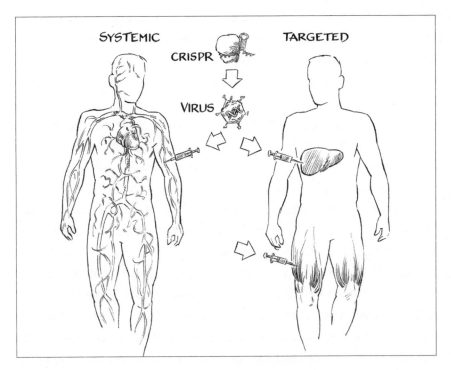

In vivo CRISPR therapy

had millions of years of evolution to perfect their craft. They're highly specialized at infecting particular types of tissues and organs, and some viruses have become relatively safe to use. Thanks to decades of genetic engineering, specialized viruses have been completely retooled so that they are still able to deliver DNA to the body—either systemically or to specific organs—but can't infect their host with anything except the therapeutic payloads that researchers give them.

One vector—the generic scientific term for a carrier of genetic information—has been an especially big asset to researchers developing in vivo gene-editing therapies: an innocuous human virus known as adeno-associated virus (AAV). AAV provokes only a mild immune response and is not known to cause any human disease. This viral vector can easily be outfitted with therapeutic genes that encode the Cas9 protein and its guide RNA, and it is highly effective at delivering its genetic material to host cells. What's more, the virus can be engineered so that it doesn't permanently incorporate its DNA into the human genome the way that other viruses do. This feature helps avoid errant DNA insertion into sensitive parts of the genome, an event that has marred other therapy efforts in the past.

Another encouraging aspect of AAV is its sheer natural diversity. By isolating different strains of the virus and then mixing and matching them in different ways, researchers have assembled a family of AAV vectors that can target cells in many different types of tissues. One strain of AAV might be best suited to deliver CRISPR to cells of the liver, while another might work best in the central nervous system, the lungs, the eyes, or the cardiac and skeletal muscles.

It's in muscles where we've seen one of the earliest and most dramatic demonstrations that CRISPR can ameliorate the ravaging effects of genetic disease in vivo. While the technique was demonstrated in a mouse model, there's every reason to think that it will be effective in human subjects as well—not least because the genetic disease it was used to treat is, sadly, prevalent in our species.

The fatal muscle-wasting disease known as Duchenne muscular dystrophy (DMD) is the most common type of muscular dystrophy in the world, inherited by roughly 1 in every 3,600 male babies. DMD patients have no symptoms at birth, but the disease appears — and progresses with devastating speed — beginning around age four. Patients suffer severe muscle degeneration and are typically confined to a wheelchair by age ten; most perish by the time they're twenty-five, overwhelmed by breathing complications and the deterioration of that most vital muscle, the heart.

Duchenne muscular dystrophy can arise from any one of several mutations in the *DMD* gene — the largest human gene known — which encodes a protein called dystrophin. This protein helps muscle cells contract, and the lack of a functional dystrophin protein is the underlying problem for DMD patients. Males are disproportionately affected; since the *DMD* gene is found on the X chromosome and males possess only one X chromosome (paired with a paternally inherited Y chromosome), a single mutated copy of *DMD* leaves them wholly devoid of healthy dystrophin. Females, however, have two X chromosomes and thus two copies of *DMD*; as long as one of the two copies is healthy, it can stave off the disease's awful symptoms. While these females are spared, however, they remain carriers of the disease and will transmit the mutated *DMD* gene to roughly half of their male offspring. (This inheritance pattern makes DMD an example of an X-linked recessive genetic disease.)

Will CRISPR be able to reverse the effects of DMD? The jury is still out — we'll need years of research and clinical trials to answer that confidently — but if recent studies in mice are any indication, there's reason to hope that in vivo therapy will do so. By the end of 2015, no fewer than *four* independent laboratories delivered CRISPR to fully grown mice suffering from muscular dystrophy and showed that the ravages of the disease could be reversed. By packaging genetic instructions for CRISPR into AAV vectors, the researchers repaired skeletal and cardiac muscle cells, either by injecting the loaded viruses into the mouse's muscles or by delivering the viruses to the same tissues through the bloodstream. They succeeded in turning on the healthy dystrophin genes, and the treated mice even showed substantial enhancement in muscle force after receiving the therapy.

I attended a presentation of these data by Eric Olson, a professor at the University of Texas Southwestern Medical Center, and was elated at the progress using in vivo CRISPR therapies. And this work makes me hopeful that it will one day be possible to treat or even cure other genetic diseases besides DMD. For instance, using a version of CRISPR programmed to edit a different gene and a version of AAV better suited to target the liver, a team at MIT used gene editing to cure mice of a genetic mutation that causes a condition known as tyrosinemia. In humans, the disease can cause an accumulation of toxic metabolites and extensive liver damage; if untreated, patients usually die before age ten. In the mouse model, however, CRISPR repaired the damaged gene and reversed the course of the disease.

AAV has also delivered CRISPR into the brains of adult mice, into their lungs, and into retinal cells in their eyes, all of which may eventually translate into therapies to treat genetic disorders like Huntington's disease, cystic fibrosis, and congenital blindness. Indeed, the first gene therapy drug approved for commercial use in the Western world uses an AAV vector, and it's possible that the first CRISPR-based gene-editing drug relying on in vivo delivery will do the same.

Yet AAV is just one of the many delivery strategies that have been developed to transport CRISPR into living cells. In the viral world alone, there's a menagerie of retooled viral Trojan horses available for use, each with its unique set of advantages and disadvantages. One example is the adenovirus, which causes the common cold among other illnesses (and assists adeno-associated viruses during infections, giving AAV their name). After gutting these adenoviruses and removing their pathogenic genes, researchers can insert a greater amount of therapeutic DNA than is allowed by AAV vectors. Lentiviruses, the most prominent example of which is HIV, have also been defused in the laboratory and converted into effective delivery vehicles. Their capacity is similar to that of AAV, but they have the ability to permanently splice their genetic material into the genome of the cells they invade. This feature is useful for basic research in the lab, but for in vivo therapies, scientists can shut off the splicing function.

And then there are in vivo delivery strategies that won't use viruses at all. Building on advances in nanotechnology — the science of fabricating submicroscopic structures — researchers are exploring the use of lipid nanoparticles to ferry CRISPR throughout the body. Resistant to degradation and easy to manufacture, these delivery vehicles also have the benefit of releasing the Cas9 protein and its guide RNA into the patient's body in a regimented way. Viruses (and their CRISPR cargo) can persist in cells for a long time, which — as I'll explain — can cause problems in the editing process, but lipid nanoparticles deliver CRISPR so that it acts quickly before being broken down by the natural recycling factories of the cell.

Beyond leading to cures for certain genetic disorders, there's one more way in which CRISPR is poised to revolutionize human health. This biotechnology is also having a game-changing impact in the study and treatment of one of the most feared diseases known to humankind: cancer.

Cancer is caused by DNA mutations, some of which are inherited and some of which are acquired over the course of one's life. So it might seem obvious that gene editing can help treat cancer, or even prevent it, by eliminating these mutations before they have a chance to do irreversible harm. But this isn't actually where CRISPR is making the biggest contribution — at least, not yet.

Rather than being a form of treatment in and of itself, CRISPR is advancing cancer care as a tool and a support system for existing therapies. It is expanding our understanding of cancer biology, and it is also accelerating immunotherapy treatments, which harness the body's own immune system to fight cancer. On both of these fronts, CRISPR is proving its worth as another weapon — one of the most powerful — in our growing arsenal in the age-old war against this fearsome disease.

Of all CRISPR's contributions to medicine, this is the one for which I feel the most personal anticipation. Even if you haven't struggled with cancer yourself, chances are you know someone whose life has been harmed or cut short by the disease. In my own family, my father's death from melanoma was a deeply affecting experience that highlighted the many challenges of dealing with such a complex disease. Cancer is one of the most common causes of death in the United States, second only to heart disease. While improvements in early diagnosis and treatments have lifted survival rates considerably over recent decades, deaths from cancer are still a devastating part of everyday life. In the United States alone, over one and a half million new cancer cases are diagnosed annually, and half a million people die from cancer every year. That's nearly two thousand deaths a day.

The DNA mutations linked to cancer are sometimes inherited; they can also arise spontaneously or be induced by tobacco use or exposures to other carcinogens. Over the past decade or so, there has been a major push to use DNA sequencing to catalog the many mutations that distinguish cancerous cells from normal, healthy cells. If these mutations can be identified, the thinking goes, then drugs can be designed to combat whatever abnormal genes are causing the malignant cells to proliferate.

But there's a problem: we simply have too much information. The critical cancer-causing mutations are buried in a vast sea of auxiliary mutations that don't directly affect disease pathology. In fact, one of the hallmarks of cancer is the increased rate at which DNA mutations creep into the genome, making it difficult to identify the mutations that are actually playing the largest role in causing tumors.

Before CRISPR, the arsenal of tools to study cancer-causing mutations was rather limited: scientists could detect and diagnose mutations in biopsies taken from patients, and they could study a small number of discrete mutations in mouse models. But now that researchers have a way to precisely replicate cancer-causing mutations — single ones, or many at a time — in a fraction of the time that was previously required and at a fraction of the cost, cancer research is poised to explode. Instead of painstakingly selecting the correctly mutated cells (an ordeal with one-in-a-million efficiencies) or breeding the desired mouse models over numerous generations (requiring years of time), scientists can use CRISPR to efficiently introduce mutations in a single pass. This capability is allowing scientists to better understand the exact genetic factors that cause cells to stop responding to the signals that normally regulate their growth.

For example, in one study from Harvard Medical School, Benjamin Ebert's team of researchers wanted to understand the genetic causes of acute myeloid leukemia, a cancer of white blood cells. By programming CRISPR to edit different genes — using a different guide RNA to match each one — they set out to knock out eight candidate genes. This kind of multiplexed gene editing would have been unthinkable before, but with CRISPR, it was straightforward. After editing the genes in blood stem cells in all different combinations and then injecting the cells back into the bloodstream of live mice, the researchers waited to see which animals came down with acute leukemia (a sort of reverse ex vivo application). Then, by cross-checking those animals with the genes that had been inactivated with CRISPR, Ebert's team deduced the exact constellation of gene mutations necessary and sufficient to promote leukemia. Experiments like these are proving invaluable for advancing human cancer research.

The ability to target many genes at once is one of CRISPR's greatest attributes. Unlike the gene-editing technologies that preceded CRISPR, the process of designing CRISPR to home in on a new twenty-letter sequence in the genome is simple enough for a high school student to master — so simple, indeed, that a computer can be programmed to do it. Scientists are now combining computer science and gene editing to probe the depths of the genome, hunting for new cancer-associated genes without any a priori information about them.

The technical details are complex, but in effect, this ultimate multi-plexing approach allows researchers to edit and knock out every single gene in the genome — in a single experiment. David Sabatini, a professor at MIT, was one of the first to pioneer such a genome-wide knockout screen. But instead of asking what gene mutations *caused* the cancer (as Ebert's team had done), Sabatini's team wanted to discover gene muta-tions that *disabled* the cancer. In other words, they wondered whether there were genes that the cancerous cells absolutely depended on for their pathogenicity and couldn't live without. In a true tour de force, Sabatini's team addressed this question for four different blood-based cancer lines and discovered a whole host of new genes that seemed to be essential for them to thrive. By identifying new genetic susceptibilities of leukemias and lymphomas, these experiments revealed promising new targets for chemotherapy drugs.

Subsequent experiments by other laboratories have revealed the weak spots of other types of cancer, among them colorectal cancer, cervical cancer, melanoma, ovarian cancer, and glioblastoma (a particularly ag-gressive cancer of the brain). Researchers have even used genome-wide knockout screens to identify new genetic factors that give cancer cells the dreadful ability to circulate in the bloodstream and invade other tis-sues, a process known as metastasis.

Advances in our basic understanding of cancer can be slow to deliver actionable information and concrete treatments, but the importance of this research can't be overstated. As medicine becomes more and more personalized, scientists and physicians confront a wealth of information that may distinguish one individual's cancer from another and that will offer hints of how to tailor treatments to match the specific biology of each particular disease. Gene-editing tools will help to make sense of this information by revealing which mutations are most predictive of cancer and which mutations might make a cancer more or less responsive to various drugs.

But while gene editing will contribute vital intelligence to our war on cancer, there are also signs that it will help us take the fight *to* cancer. Its most promising role in this regard is as a support system for a form of treatment that's gotten a lot of attention in recent years: immunotherapy.

A revolutionary form of cancer treatment, immunotherapy is a departure from the three main types — surgery, radiation, and chemotherapy — that doctors have historically employed. Unlike these older approaches, cancer immunotherapy aims to use a patient's own immune system to hunt down and destroy dangerous cells. In a complete paradigm shift, immunotherapy targets not the cancer, but the patient's own body, empowering it to fight cancer on its own.

The core idea behind cancer immunotherapy is to tweak the human immune system, specifically the T cells, its primary foot soldiers. By rewiring these cells to recognize the molecular markers of cancer, scientists can help T cells mount an immune response to eradicate cancerous cells. The challenge is figuring out how to unleash the full potential of T cells.

One promising development involves checkpoint inhibitors, drugs that shut off the brakes that usually restrain the immune response against cancerous cells. Another strategy involves the manufacture of genetically engineered T cells that are precisely designed to target a patient's unique cancer. This process — yet another example of ex vivo therapy — is known as adoptive cell transfer, and it's in this mode of immunotherapy that gene editing enters the picture.

The basic goal of adoptive cell transfer, or ACT, is to help T cells better target cancer cells. T cells are endowed with a new gene that produces a protein, called a receptor, that is fine-tuned to recognize and target the molecular markers of cancer. But there's a problem: T cells already possess a natural receptor gene, and having multiple such genes simultaneously causes molecular chaos. Researchers can now avoid this problem by using CRISPR to knock out the original receptor gene, clearing valuable space for the cancer-seeking receptor gene. Other kinds of gene edits can then be layered on top of the first to make the engineered T cells even more potent.

It seems likely that gene editing will go a step further and transform cancer immunotherapy into more of an off-the-shelf treatment, where a single batch of engineered T cells, designed for a specific type of cancer, could be given universally to all patients suffering from that pathology. Clinical trials are currently under way to test this mode of cell transfer, and a moving story from late 2015 hints at its amazing potential. In fact, the subject of this story, Layla Richards, is the first human whose life was saved by therapeutic gene editing.

Layla was a one-year-old London patient suffering from acute lymphoblastic leukemia, the most common type of childhood cancer. Her doctors had to admit that hers was one of the most aggressive leukemias they had ever seen. Although some 98 percent of children go into remission after starting treatment, Layla's condition had not improved despite chemotherapy, bone marrow transplantation, and an antibody-based pharmaceutical drug. Transplanting Layla's own engineered T cells back into her body wasn't an option either; her immune system was so weakened by the leukemia — a disease that, after all, affects the same white blood cells needed for a healthy immune system — that she no longer had enough T cells for doctors to extract.

Layla's situation looked bleak, and her doctors had already offered palliative care to ease her death. But then, at the last minute, another option presented itself.

The very same hospital that was taking care of Layla housed a facility that was editing T cells using TALENs, one of the precursor technologies to CRISPR. The cells were being prepared by a French biotechnology company called Cellectis to be used for ACT in clinical trials. After receiving consent from her parents and Cellectis, Layla's doctors tried out these untested cells for the first time ever on a human patient, which is permitted under what is known as compassionate use.

The T cells Layla received were special for a few reasons. First, they contained a new receptor gene specifically designed to target any cells that contained the molecular signature of leukemia. Second, the T cells had been gene-edited to prevent them from mounting an immune response against Layla's own cells, as they would otherwise have done (the donor and recipient in this case were not immunocompatible). Finally, the T cells contained edits to another gene that gave them a kind of invisibility cloak so that they could survive longer in Layla's body.

In the weeks that followed the cell transfer, a miraculous transformation took place in the one-year-old girl; her leukemia began responding to the edited T cells. When her health improved enough, Layla underwent another bone marrow transplant, and within months, the cancer was in complete remission. What had begun as a major gamble — attempting a treatment that had up until that point been tested only in mice — ended up a resounding success and a major endorsement for using gene editing to further immunotherapy.

Thanks to Layla's case and others, CRISPR-based therapeutics companies have already struck major deals with cancer immunotherapy companies to combine their respective platforms. Editas Medicine has an exclusive multimillion-dollar license with Juno Therapeutics to develop T cell therapies, and Intellia Therapeutics has partnered with the major health-care firm Novartis to similarly pursue cancer immunotherapy. The National Institutes of Health has approved the first U.S. clinical trial involving CRISPR-edited cells from scientists at the University of Pennsylvania, and in October 2016, a group of Chinese scientists at Sichuan University became the first to inject human patients with cells that had been modified using CRISPR. These laudable endeavors will help bring the benefits of gene editing to patients in need.

I sincerely hope that Layla's story is someday unremarkable, just another example of a human life saved by a breakthrough in gene editing. Certainly we are inching our way closer to that bright future. Before we get there, however, we'll have to solve one major problem with gene editing. Until it's fixed, this issue, which has to do with the accuracy of CRISPR's edits, will mean that stories like Layla's are the exception rather than the rule.

CRISPR — at least, the original version of it found in bacteria — is not an entirely error-free method of targeting and cutting DNA. This was clear from the very first experiments we conducted with CRISPR in my lab.

After figuring out the basic function of CRISPR, Martin set out to gauge the DNA-cutting accuracy of the Cas9 enzyme and its guide RNA. This tiny homing missile seemed to be able to find and attack any sequences of DNA that matched its guide RNA, tracking them down with impressive exactness. But were there limits to how accurate it could be? Could CRISPR truly discriminate between one twenty-letter sequence — the one matching the RNA — and other sequences that might differ by only one or two letters of DNA? If we were to have any hope of transforming this bacterial defense system into a gene-editing tool that would be safe enough to use on humans, we'd first have to answer this question.

Martin found that, when he unleashed this CRISPR machinery on DNA sequences that had been intentionally misspelled so that certain letters no longer matched the RNA, the Cas9 enzyme would in some cases still cut the DNA. Unlike the precision of a search function on a computer, where the query for the word *affect* will never turn up results for *effect*, it seemed CRISPR could make occasional mistakes and confuse one letter of DNA for another.

Later, my lab repeated these experiments in much greater depth together with a team from Harvard University led by David Liu. We exhaustively tested different DNA mutations to determine which kinds of off-target sequences (that is, those *not* matching the guide RNA) were similar enough to the intended sequence that *did* match the guide

RNA (that is, the on-target sequence) to still get recognized and cut by CRISPR. Other labs also conducted similar experiments inside cells to show that errant CRISPR cutting could lead to permanent edits at unintended sites.

To be sure, virtually all medical drugs have some kind of off-target activity, and as long as the intended on-target benefits outweigh those risks, physicians and regulators are generally pretty forgiving. For instance, antibiotics kill off both pathogenic bacterial strains and beneficial strains, and chemotherapy drugs kill off both cancerous cells and healthy cells. The challenge is ultimately one of specificity: developing a drug that is so closely matched to its intended target that a few atoms out of place will weaken the interaction enough to prevent the drug from causing unintended effects.

Usually, some degree of off-target activity is unavoidable, which is why every drug on the market carries warnings of side effects — but when it comes to gene editing, side effects could be especially dangerous. After all, the side effects of a medication typically cease once a patient stops taking the drug. With gene editing, however, any off-target DNA sequence, once edited, is irreversibly changed. Not only will unintended edits to the DNA be permanent, they will also be copied into every cell that descends from the first one. And although most random edits are unlikely to damage the cell, if we've learned anything from certain diseases and cancers, it's that even a single mutation can be enough to wreak havoc on an organism.

Luckily, off-target edits made by CRISPR, like other gene-editing technologies, tend to be fairly predictable, since they affect only the DNA sequences that are most similar to the matching guide RNA. If CRISPR is programmed to target a twenty-letter sequence in gene X, but gene Y has a similar DNA sequence that differs in only one letter, there is a finite probability that CRISPR will introduce edits in both genes. The less closely the two sequences mirror each other, the lower the likelihood of off-target mutations.

Researchers have already begun to find ways around this potential problem. Multiple laboratories have written computer algorithms that will automatically probe the three-billion-letter human genome to see how many other regions have sequences similar to the one a scientist wants to edit. If the number of potential off-target DNA sequences is too high, the researcher, aided by the algorithm, can simply select a new region to target. (In many cases, scientists can edit the same gene by choosing from a number of closely spaced DNA sequences.) The problem with this approach, though, is that the computer algorithm, no matter how well designed, may not always successfully predict off-target edits.

These "known unknowns" have led researchers to adopt a second strategy: assume complete ignorance. They assume that every version of CRISPR will inevitably have unpredictable off-target effects, and that the only way to detect them is to just try the experiment first and then go hunting for new mutations in places where they shouldn't be. Instead of computational prediction, this strategy simply involves empirical testing. Before ever selecting a DNA sequence to edit in patients, scientists will exhaustively test a bunch of related DNA sequences in cultured cells, determine which ones have the least number of off-target effects, and only then — once they have the winner — will they proceed to clinical trials.

There's a third strategy for avoiding potential off-target mutations, one with which scientists have already made great headway: engineering CRISPR to be more discriminating in how it recognizes the target DNA. For example, scientists have successfully expanded the sequence of DNA that CRISPR has to recognize, minimizing the chances of an unlucky mismatch — a strategy not unlike increasing the length of a computer password to reduce the likelihood that someone can guess it. By simply tweaking the natural Cas9 protein in a few different places — swapping out one amino acid for another — researchers, including Keith Joung from Harvard Medical School and Feng Zhang from MIT, have developed higher-fidelity versions of CRISPR that are less prone to off-target gene editing than the version nature evolved on its own.

Finally, the dosage of CRISPR affects the likelihood of the genome being riddled with unintended mutations. In general, the more Cas9 and guide RNA a cell gets and the longer those molecules hang around, the more likely it is that CRISPR will find slightly related but mismatched sequences and introduce off-target edits. The trick is to deliver just enough CRISPR into cells so that the right DNA target sequence gets edited, but no more than that.

By fine-tuning these tactics in the laboratory, researchers continue working to ensure that CRISPR can be used safely in patients. If their success so far is any indication, it won't be long before the accuracy of this molecular machine is sufficient to allow it to move from the laboratory to the clinic.

The CRISPR technology was born just a handful of years ago, but it's becoming difficult to find diseases for which it *hasn't* been mentioned as a possible therapy. Beyond cancer, HIV, and the genetic disorders discussed thus far, a quick survey of the published scientific literature reveals a growing list of diseases for which potential genetic cures have been developed with CRISPR: achondroplasia (dwarfism), chronic granulomatous disease, Alzheimer's disease, congenital hearing loss, amyotrophic lateral sclerosis (ALS), high cholesterol, diabetes, Tay-Sachs, skin disorders, fragile X syndrome, and even infertility. In virtually all cases where a particular mutation or defective DNA sequence can be linked to a pathology, CRISPR can in principle reverse the mutation or replace the damaged gene with a healthy sequence.

Because of the ease with which it can find and repair any sequence of DNA, CRISPR has often been hailed as the breakthrough that will finally eliminate disease. Yet things are never quite so simple. There are all sorts of disorders — from autism to heart disease — that don't show significant genetic causation or are caused by a complex combination of genetic variants and environmental factors. In these cases, gene editing may be of more limited use. Then, too, while gene editing is capable of repairing DNA in cultured human cells, it will be years before its efficacy is (or is not) demonstrated in human patients, and the few clinical successes that have been achieved so far with cancer immunotherapy and HIV might or might not be accurate predictors of other successes to come.

Previous genetic-engineering technologies, including gene therapies and RNA interference, were similarly extolled as pivotal advances that would completely transform medicine, yet hundreds of clinical trials have thrown quite a bit of cold water on that enthusiasm. That's not to say that we're heading for the same sort of rude awakening with gene editing, just that it's important to temper the excitement with realistic expectations, methodical research, and meticulous clinical trials. Only then can we ensure that the first wave of CRISPR-based therapeutics will have the best chance of success and the least risk of dangerous side effects.

As of this writing, the field of gene-editing-based therapies is expanding at a frantic pace, both in the academic and commercial realms. New studies surface at a rate of more than five per day, on average, and investors have poured well over a billion dollars into the various startup companies that are pursuing CRISPR-based biotechnology tools and medical therapeutics.

I am extremely excited and enthusiastic about virtually all the phenomenal progress being made with CRISPR — save for the advancements on one front. I think we should refrain from using CRISPR technology to permanently alter the genomes of future generations of human beings, at least until we've given much more thought to the issues that editing germ cells will raise. Until we have a better understanding of all the attendant safety and ethical issues, and until we have given a broader range of stakeholders the opportunity to join the discussion, scientists would do well to leave the germline alone. But, really, whether we'll ever have the intellectual and moral capacity to guide our own genetic destiny is an open question — one that has been on my mind since I began to realize what CRISPR was capable of. For this reason and others, I've come to see a clear boundary between the procedures described in this chapter and those involved in germline editing. We should think twice before crossing that line. And then we should think again.

7

THE RECKONING

IN THE SPRING OF 2014, roughly a year before my first Davos meeting, I got a taste of what would soon become an international effort to shape the future of CRISPR.

It had been less than two years since the publication of our paper in *Science* describing how CRISPR could be harnessed for gene editing, but news of this technology had already spread through the scientific community — and beyond. Popular excitement about CRISPR had begun to grow, thanks to enthusiastic descriptions of gene-editing research in the mainstream media. As research using the CRISPR technology continued to accelerate, many scientists tried to keep their focus in the laboratory — advancing gene-editing methods themselves and using these methods in new ways — without getting swept into a more public discussion.

Like these colleagues, I had continued to explore and develop CRISPR, working with my academic lab at Berkeley while also devoting an increasing amount of time to better understand the challenges of using gene editing for human therapeutics. Such work was ongoing in numerous academic labs and was also getting under way at several startup biotechnology companies. It was exhilarating to be part of what felt like a massive collective effort to uncover the workings of the CRISPR technology and unlock its vast potential to manipulate genetic information inside cells. Mostly I felt excited and hopeful that our efforts would bring about positive developments in fields ranging from agriculture to medicine. But occasionally, I found myself lying awake in the wee hours of the night wondering about people outside of academia who were also taking a keen interest in this burgeoning field, and not always for the best reasons.

It was around this time that my coauthor, Sam Sternberg, then a PhD student in my lab, received an e-mail from an entrepreneur whom I'll call Christina. She wanted to know if Sam would be interested in being a part of her new company, which somehow involved CRISPR, and she asked to meet him so she could pitch her business idea.

On the face of it, Christina's note wasn't surprising. Given the pace at which CRISPR had been developed and disseminated, and given its increasingly obvious potential to disrupt so many sectors of the biotechnology market, every week seemed to bring word of yet another new company, product, or licensing deal related to gene editing. But as Sam would soon discover, Christina's venture was different — very different.

Sam didn't really know what to expect when he met Christina for dinner at an upscale Mexican restaurant near campus, but he was nonetheless caught off guard by their conversation. Her e-mail had been vague, but in person Christina spoke freely about what she hoped to do with the technology Sam was helping to develop.

Speaking passionately over cocktails, Christina told Sam that she hoped to offer some lucky couple the first healthy "CRISPR baby." The child, she explained, would be produced in the lab using in vitro fertilization, but it would have special features: customized DNA mutations, installed via CRISPR, to eliminate any possibility of genetic disease. While trying to entice him to come on board as a scientist, Christina assured him that her company planned to introduce only prophylactic genetic modifications in human embryos; if he wanted to be involved, he needn't worry about making any mutations that weren't necessary to ensure the health of the unborn child.

Christina didn't have to explain to Sam how the procedure would work or how easy it would be. To edit the human genome in the way she

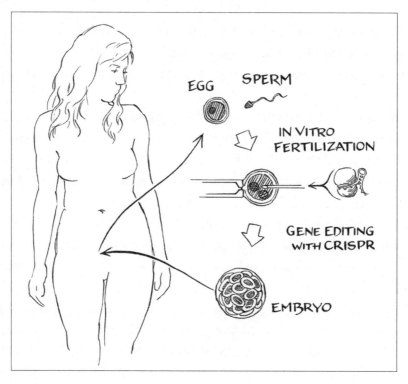

The prospect of editing genes in human embryos

was suggesting, a clinician would need only techniques that were already well understood by this time: generation of an embryo in vitro from the would-be parents' egg and sperm cells, injection of preprogrammed CRISPR molecules to edit the embryo's genome, and implantation of the edited embryo into the mother's uterus. Nature would take care of the rest.

Sam excused himself before dessert; he'd heard enough. Despite Christina's assurances, he came away from their conversation rattled. She was, Sam sensed, obsessed with the power and possibilities of CRISPR. As he told me later, he'd perceived a Promethean glint in her eyes and suspected she had in mind other, bolder genetic enhancements in addition to the well-intentioned genetic changes she'd described.

Had this conversation occurred just a few years earlier, Sam and I would have dismissed Christina's proposal as pure fantasy. Sure, genetically modified humans made for great science fiction, and they were a fertile subject for philosophical and ethical musings on the possibility of human "self-evolution." But unless the *Homo sapiens* genome suddenly became as easy to manipulate as the genome of a laboratory bacterium like *E. coli*, there was little chance of anyone pursuing such Frankenstein schemes anytime soon.

Now, we could no longer laugh off this kind of speculation. Making the human genome as easily manipulable as that of a bacterium was, after all, *precisely* what CRISPR had accomplished. Just a month before Sam's meeting with Christina, in fact, the first monkeys were born with genomes that had been rewritten through precision gene editing, bringing the steady march of CRISPR research right to *Homo sapiens'* evolutionary front door. In light of this development with primates and the number of CRISPR-modified animal species — from worms to goats — that had preceded them, it seemed only a matter of time before humans were added to the growing list of creatures whose genomes had been reworked.

I felt keenly aware of, and apprehensive about, this possibility. While I couldn't gainsay the many overwhelmingly positive effects that gene editing would have on our world — allowing us to better understand human genetics, produce food more sustainably, and treat victims of devastating genetic diseases — I was growing anxious about other uses to which CRISPR could conceivably be put. Had our discovery made gene editing *too* easy? Were scientists rushing too haphazardly into new areas of research without stopping to think about whether their experiments were justified or what their effects might be? Could CRISPR be misused or abused, particularly where the human genome was concerned?

I was especially beginning to worry if, someday soon, scientists would attempt to alter the human genome in a heritable way, not to treat a disease in a living patient, but to eliminate the *prospect* of disease in a child who hadn't yet been born or even conceived. This was, after all, exactly what Christina had proposed to Sam. Even if she didn't accomplish it, who was to say that someone else wouldn't?

This possibility gnawed at me. Humans had never before had a tool like CRISPR, and it had the potential to turn not only living people's genomes but also all future genomes into a collective palimpsest upon which any bit of genetic code could be erased and overwritten depending on the whims of the generation doing the editing. What's more, encounters such as Sam's meeting with Christina were forcing me to acknowledge that not everyone shared my trepidation about the prospect of scientists rewriting the DNA of future human beings without fully appreciating the consequences. Somebody was inevitably going to use CRISPR in a human embryo — whether to eradicate the sickle cell trait in a single family's germline or to make nonmedical enhancements — and it might well change the course of our species' history in the long run, in ways that were impossible to foretell.

The question, I was beginning to realize, was not *if* gene editing would be used to alter DNA in human germ cells but rather *when,* and *how.* It was also becoming clear to me that, if I wanted to have a say in when and how CRISPR would be used to change the genetic makeup of future humans, I would first have to understand exactly how much of a break from previous scientific accomplishments germline editing would be. What sorts of interventions in the human germline had previously been achieved, and how had they been tolerated? What were the goals of these earlier interventions? And what had my predecessors — in particular, the scientific luminaries of generations past — had to say about human germline manipulation, the prospect of which alarmed me so much?

It's not as if the debate over modifying the human germline began only when CRISPR came along. Far from it. Back when the earliest hints of gene editing were emerging, physicians in reproductive medicine were already selecting certain embryos over others for establishing pregnancies, thus making choices about what genetics would be propagated to subsequent generations. And for even longer, practitioners and observers of science have been disturbed by the notion that humans might someday be the primary authors of their own genetic constitutions.

Once the role of DNA in encoding genetic information had been proved, researchers began to appreciate the power of rationally manipulating genetic code, even though the tools for doing so didn't yet exist. Marshall Nirenberg, one of the biologists responsible for cracking the genetic code in the 1960s (a feat for which he was awarded the Nobel Prize in Physiology or Medicine), wrote in 1967 of man's "power to shape his own biologic destiny. Such power," he observed, "can be used wisely or unwisely, for the betterment or detriment of mankind." Aware that such a capability should not lie in the hands of scientists alone, Nirenberg continued, "Decisions concerning the application of this knowledge must ultimately be made by society, and only an informed society can make such decisions wisely."

Not all scientists were so restrained. Writing in the *American Scientist* just a few years later, Robert Sinsheimer, then a professor of biophysics at Caltech, described human genetic modification as "potentially one of the most important concepts to arise in the history of mankind . . . For the first time in all time a living creature understands its origin and can undertake to design its future." Sinsheimer scoffed at critics who argued that genetic engineering was simply a modern version of the timeless but futile dream to perfect mankind: "Man is all too clearly an imperfect, a flawed creature. Considering his evolution, it is hardly likely that he could be otherwise . . . We now glimpse another route—the chance to ease the internal strains and heal the internal flaws directly—to carry on and consciously to perfect, far beyond our present vision, this remarkable product of two billion years of evolution."

Within two decades of the publication of Sinsheimer's essay, scientists were rapidly mapping the route to perfection he had been able only to glimpse in the late 1960s. By the beginning of the 1990s, gene therapy trials were under way with human patients—and while it was clear that accurately manipulating the human germline wouldn't be feasible even with this relatively advanced technology, that didn't stop researchers from wringing their hands about the possibility. French Anderson, the scientist who led those first clinical trials, was outspoken about the hazards and ethical arguments against using gene therapy for enhancement purposes, whether in somatic cells or in the germline. Above all, he questioned whether any scientist could wield this newfound power responsibly or if instead the scientist "might be like the young boy who loves to take things apart. He is bright enough to disassemble a watch, and maybe even bright enough to get it back together again so that it works. But what if he tries to 'improve' it? Maybe put on bigger hands so that the time can be read more easily. But if the hands are too heavy for the mechanism, the watch will run slowly, erratically, or not at all . . . Attempts on his part to improve the watch will probably only harm it."

Despite the warnings of leading scientists like Anderson, the idea of altering or refining our genetic makeup continued to galvanize some biologists throughout the last decade of the twentieth century. The excitement of these women and men was stoked by ongoing research and development into human gene therapy as well as by seminal advances in three major areas: fertility research, animal studies, and human genetics.

Back then, any scientists dreaming of someday "improving" on the genetic makeup of the human race and searching for inspiration had to look no further than recent advances in the treatment of infertility. The birth of Louise Brown in 1978, the world's first "test-tube baby," was a watershed moment for reproductive biology, proving that human procreation could be reduced to simple laboratory procedures: the mixing of purified eggs and sperm in a petri dish, the fostering of a zygote as it grew into a multicellular embryo, and the implantation of that embryo in the female womb. In vitro fertilization, or IVF, enabled parents with various forms of infertility to produce genetically related children while also opening the door to other manipulations that could eventually be performed on the early-stage embryo during its growth in the laboratory. After all, if a human life could be created in a petri dish, the same type of sterile environment where gene-editing technologies were being developed, it was conceivable that the two methods would someday converge. Research aimed at circumventing infertility had inadvertently refined a procedure that would become integral to future discussions of germline manipulation.

Animal research, too, encouraged scientists who thought that human germline editing was almost within reach. Over the last few decades of the twentieth century, scientists had devised more and more ingenious ways of engineering animal genomes, from cloning to virus-based gene addition to the earliest uses of precision gene editing. By the 1990s, it had become fairly routine to generate mouse models of human disease by modifying specific genes in the mouse germline; although the exact procedure couldn't be used on humans, it set the stage for inventions like ZFN and CRISPR, which transformed the formerly crude method of germline gene editing in mice into a streamlined, exact, and highly optimized method that was much better suited to human subjects. The decade also witnessed the first successful cloning of a mammal with the famous birth of Dolly the sheep in 1996. By transferring the nucleus (with all its DNA) of a somatic cell taken from an adult sheep into a recipient egg cell whose nucleus had been removed, stimulating the hybrid cell to begin dividing, and then implanting the resulting embryo in a surrogate mother, Ian Wilmut and his colleagues in Scotland produced a ewe whose genome was a perfect copy of the donor.

IVF and cloning were huge technical breakthroughs that helped lay the groundwork for germline modification. Not only did they show that scientists could generate a viable embryo in the lab by mixing egg and sperm, they also revealed that the embryos could be created using genetic information from a single animal. The feat sent regulators worldwide scrambling to enact legislation that would prohibit the reproductive cloning of humans. As it turned out, cloning mammals turned out to be so technically difficult that few laboratories in the world were capable of attempting it. Thus, unlike CRISPR, the technology of somatic cell nuclear transfer was effectively self-limiting due to the extensive expertise it required.

Finally, enthusiasm for making changes to the DNA of future humans was a natural outgrowth of breakthroughs in human genetics, especially the sequencing of the human genome. This incredible development made many people think that geneticists would soon be able to find the root causes of once-mysterious diseases as well as the genetic code for a much broader range of human phenotypes, from physical traits to cognitive ones. Once we fully understood the genetic factors that determine human health and performance, we might be able to select for — or perhaps even engineer — embryos with a genetic composition different than that of their parents. *Better* than that of their parents.

Or so some scientists hoped. I, for one, was skeptical about what I saw as blind optimism in this pre-CRISPR era, with some gushing about the possibilities of reshaping the germline without pausing to consider the consequences. Would this sort of procedure really be able to safely rid all of a person's descendants of a genetic illness, or would it have side effects that we could not foresee? It seemed impossible to ever conduct experiments that would provide answers to such questions. And even if it could be done safely, would doctors and their patients really restrict it to medical operations, or would they cross the line by making nonessential modifications? Although at the time I hadn't given all that much thought to these questions, they nonetheless nagged at me whenever the subject of the germline came up.

In 1998, the growing excitement — or unease, depending on where on the continuum you found yourself — over germline modification prompted two scientists, John Campbell and Gregory Stock, to organize one of the first symposia on the topic at the University of California, Los Angeles. Called Engineering the Human Germline, the meeting featured talks from some of the foremost researchers in the field, including French

Anderson, the gene therapy pioneer; Mario Capecchi, one of the fathers of early gene editing; and James Watson, co-discoverer of the structure of DNA. Although I wasn't in attendance — back then, my research was still focused on questions like how tiny RNA molecules folded into elaborate three-dimensional structures — records of the conference helped reassure me years later that I wasn't alone in worrying about tampering with the human germline and that my concerns were by no means new.

At the time of the UCLA conference, its participants were wrestling with many of the same concerns about germline modification that have resurfaced in recent years with the advent of CRISPR, issues such as consent, inequality, access, and unintended consequences for future generations. Like many concerned scientists today, these researchers grappled with the thorny question of whether scientists would be transgressing natural or divine laws by changing the human germline and whether such efforts would constitute eugenics, a fallacious early-twentieth-century set of beliefs and practices that have since been thoroughly repudiated by mainstream science. But in addition to, or perhaps in spite of, these weighty ethical considerations, the participants in the 1998 symposium were clearly buoyed by extreme optimism at the possibility of using the latest scientific breakthroughs to improve humanity. Panel discussions focused on topics such as eradicating disease, avoiding serious genetic defects, and generally improving on the natural course of evolution — which, attendees argued, could be so cruel as to justify some sort of intervention.

A report authored a few years later by the American Association for the Advancement of Science on the topic of human inheritable genetic modification was considerably more restrained. The working group concluded that germline interventions couldn't (yet) be carried out safely or responsibly, that the ethical concerns were serious, and that the risks of germline modification being used for enhancement purposes were especially problematic. A few years later, the Genetics and Public Policy Center reached similar conclusions, while also acknowledging that consumer demand for certain uses was likely to evolve if scientists developed viable procedures.

In addition to these conferences and reports, another event foreshadowed some of the goals — and the controversies — of germline modification that would gain new urgency with the birth of CRISPR. This was the advent of a medical procedure that allowed parents to choose, albeit in a limited way, the genetic material that their children would inherit.

Once the technique of in vitro fertilization transformed the act of conception into a rather simple laboratory procedure, it became feasible to subject early-stage human embryos to DNA sequence analysis just like any other biological sample. Since each parent passes down only 50 percent of his or her DNA to offspring, the particular constellation of chromosomes and genes that a child inherits is essentially random. But the ability to generate multiple embryos in the laboratory using multiple eggs and sperm changed all that. Instead of implanting random embryos into the mother, IVF doctors could first analyze the DNA of candidate embryos to make sure they were selecting ones with the healthiest genomes — a practice that's come to be known as preimplantation genetic diagnosis, or PGD.

Of course, prenatal genetic testing exists for embryos that arise from sexual reproduction as well, and it is increasingly practiced today. Amniocentesis or a simple blood sample taken from the mother (which harbors trace amounts of the fetus's DNA) can reveal chromosomal abnormalities like Down syndrome and even specific disease-causing gene mutations. But there are still ethical issues to consider. After all, if prenatal testing indicates that a fetus suffers from damaging genetic defects, there are typically only two options: proceed with the pregnancy or terminate it. Unsurprisingly, given the controversy surrounding selective abortion, the use of this sort of testing has been the source of vigorous debate.

Preimplantation genetic diagnosis avoids difficult issues like these by making embryo selection possible before a pregnancy is even established (though it also requires fertilization to take place in vitro, which is costly and involves invasive egg retrieval from the mother). PGD still suffers from technical challenges, but on the whole, it has been effective in preventing the birth of children with certain kinds of genetic conditions, and it has become an attractive option for parents who are already considering IVF because of fertility problems. Yet while this technique avoids the ethical conundrum of abortion, it has some heavy philosophical baggage of its own.

In its earliest implementations, PGD was used for gender selection, albeit for medical reasons; diseases linked to mutations on the X chromosome, known as X-linked diseases, could be specifically avoided if female embryos were selected. But despite scientists' good intentions, many observers and regulators simply couldn't abide the idea that PGD allowed parents to decide whether to have a boy or a girl — especially since, in many countries, female children are considered less desirable than males. Today, the use of preimplantation genetic diagnosis for gender selection is illegal in many countries (including India and China) or permitted only to avoid X-linked diseases (as in Great Britain). But it's legal in the United States, where many fertility clinics offer it as a reproductive option to parents without requiring any urgent medical reason.

PGD has also been used for other controversial purposes, such as the birth of so-called savior siblings, destined from the moment of implantation not only to live their own lives, but also to serve as organ or cell donors for a sibling. And in the future, parents may be offered the option of selecting for traits that go beyond disease susceptibility and gender and cross into areas like behavior, physical appearance, or even intelligence. The list of known associations between certain gene variants and a diverse list of traits continues to grow, and as the PGD technology improves further, what's to stop fertility clinics from consulting this genetic information so they can offer their consumers even more choices when it comes to selecting the most desirable or "best" embryos?

The implications of this kind of genetic testing are extreme, yet it's not even the latest or most advanced technology associated with assisted reproduction. That distinction goes to mitochondrial replacement therapy, colloquially known as three-parent IVF. Paradoxically, any babies born from this procedure contain DNA from not two, but three parents: one father and two mothers. This therapy — which involves transferring the nucleus of one egg cell into another egg cell whose nucleus has been removed — aims to save unborn children from an otherwise unavoidable class of genetic conditions called mitochondrial diseases. The second egg cell contains no nucleus but does have mitochondria, which house a small portion of the human genome, so this procedure creates an individual who bears genetic relatedness to three parents: the mother who contributed the nuclear genome (and who is likely to raise the child); the mother who contributed the enucleated egg cell and its mitochondrial genome (a small but essential collection of genes); and the father who contributed the sperm and the second copy of the nuclear genome.

Mitochondrial replacement therapy has been shown to work with mice and nonhuman primates, and has already been performed on human eggs. There's still controversy about its safety, but clinical applications are on the horizon. The advisory committee that oversees fertility research and treatments in the United Kingdom endorsed mitochondrial replacement therapy in a 2014 report, and after parliamentary approval in 2015, the UK became the first country in the world to approve regulations permitting its clinical use. The United States may not be far behind; in early 2016, the National Academies of Sciences, Engineering, and Medicine similarly recommended that the Food and Drug Administration approve future trials of three-parent IVF.

Procedures like PGD and three-parent IVF demonstrate that the scientific and medical communities are willing to push the ethical envelope in order to enable parents to have healthy children. Even three-parent IVF, which is technically very similar to reproductive cloning in some regards, has come under relatively little philosophical or regulatory scrutiny compared to that other, much more controversial technique. And three-parent IVF would permanently alter the human genome, changing the germline in ways that would be passed on to future generations in perpetuity. Regulators have nevertheless greenlighted this reproductive therapy.

Reading about these cases, I had to ask myself: Would regulators and researchers be just as comfortable using CRISPR to make heritable changes to the human genome, given that its power is so much greater than these earlier technologies? When fertility doctors eventually realize that they have the ability to enhance embryos' genomes with many, many more gene variants than could be provided by any given set of parents, will they really pause to reflect on the possible consequences? Or will they rush to make use of this newfound power, blindly grasping a genetic tool that, wielded in the dark, cannot be fully controlled?

I wasn't used to asking myself these sorts of questions in my day-to-day life as a professor and biochemical researcher. Although I recall writing on my application to graduate school that I was interested in scientific communication, in truth I much preferred working in the lab and trying new experiments to thinking about the theoretical, long-term implications of my research and trying to explain them to nonscientists. And as I got more deeply involved in my field, I spent increasing amounts of time talking with specialists and less time talking to people outside my immediate circle of experts. In this way, I fell into a common trap; scientists, like anyone else, feel most comfortable when surrounded by others like themselves, people who speak the same language and worry about the same issues, big and small.

Two years after my colleagues and I published the article that described CRISPR as a new gene-editing platform, though, I was finding it impossible to ignore these big-picture questions and stay inside my familiar scientific bubble. As scientists used CRISPR to edit the genes of more and more animals, and as they continued to expand the tool's capabilities, I realized it would not be long before researchers somewhere tested CRISPR on human eggs, sperm, or embryos with the goal of permanently rewriting the genome of future individuals. But incredibly, no one was discussing this possibility. Instead, the gene-editing revolution was unfolding behind the backs of the people whom it would affect. Even as the CRISPR field was exploding, no one outside of my circle of colleagues seemed to know about it or understand what was coming. Eventually, the disconnect this created between my professional life and my personal life became profound. By day I compared notes with specialists, by night I dined with neighbors and chatted with PTA parents, and all the while I marveled over how little the denizens of these two worlds seemed to know about each other. Thus, while the UK authorities were openly deliberating over mitochondrial replacement therapy, I was privately struggling over whether I could avoid the ethical storm brewing around this technology I had helped create.

It's not that I was categorically opposed to the idea of scientists and physicians using gene editing to introduce heritable changes into the human genome. To be sure, there were numerous philosophical, practical, and safety issues — many of which I'll cover in the next chapter — that deserved in-depth discussion and vigorous debate, but none of these constituted a reason to absolutely forbid this use of the technology. I was far more concerned about two other, more concrete hazards: first, that through a series of reckless, poorly conceived experiments, scientists would prematurely implement CRISPR without proper oversight or consideration of the risks, and second, that by virtue of being so effective and easy to use, CRISPR might be abused or employed for nefarious purposes.

It was hard to know what such misuses might be and who might be committing them. Even in the spring of 2014, before I had a chance to consider these issues deeply, my subconscious was offering up answers in the form of nightmares — one of which I alluded to in the opening pages of this book.

In this particular dream, a colleague approached me and asked if I would be willing to teach somebody how the gene-editing technology worked. I followed my colleague into a room to meet this person and was shocked to see Adolf Hitler, in the flesh, seated in front of me. He had a pig face (perhaps because I had spent so much time thinking about the humanized pig genome that was being rewritten with CRISPR around this time), and he was meticulously prepared for our meeting with pen and paper, ready to take notes. Fixing his eyes on me with keen interest, he said, "I want to understand the uses and implications of this amazing technology you've developed."

His terrifying appearance and sinister request were enough to jolt me awake. As I lay in the dark, my heart racing, I couldn't escape the awful premonition with which the dream had left me. The ability to refashion the human genome was a truly incredible power, one that could be devastating if it fell into the wrong hands. The thought frightened me even more because, by this point, CRISPR had been widely disseminated to users around the globe. Tens of thousands of CRISPR-related tools had already been shipped to dozens of countries, and the knowledge and protocols needed to create designer mutations in mammals — at least in mice and monkeys — had been described in great detail in numerous published articles. To make matters worse, CRISPR wasn't employed only by the hundreds of academic and commercial research labs worldwide; it was also sold online to any consumer with a hundred dollars. Sure, these DIY CRISPR kits were designed to modify only bacterial and yeast genes, but the technique was simple enough, and academic experiments with animal genomes had become so routine, that it wasn't hard to imagine biohackers messing with more complex genetic systems — up to and including our own.

What had we done? Emmanuelle and I, and our collaborators, had imagined that CRISPR technology could save lives by helping to cure genetic disease. Yet as I thought about it now, I could scarcely begin to conceive of all of the ways in which our hard work might be perverted.

Overwhelmed by how fast everything was moving and by how quickly it seemed it could all go wrong, I began to feel a bit like Dr. Frankenstein. Had I created a monster?

As if my mind weren't occupied enough with these unsettling thoughts, I found myself worrying about yet another possibility: that scientists wouldn't conduct their research transparently. Science, after all, does not happen in a vacuum. This is especially true for the applied sciences, in which breakthroughs often have a direct impact on society. I strongly believe that scientists working in this field have a responsibility to conduct their research openly, to educate the public about their work, and to engage in collective discussions about the possible risks, benefits, and ramifications of their experiments *before* conducting any that might cross the Rubicon, so to speak.

In the case of CRISPR, it seemed clear that public discussion was falling far behind the breakneck pace of scientific research. I wondered if there might be a backlash if experiments on humans were attempted before we could have an open deliberation about gene editing. And it seemed possible that such a backlash could damage or delay more urgent and uncontroversial therapeutic applications of CRISPR, such as the treatment of genetic diseases in adult patients. Increasingly concerned by these prospects, I fumbled for clues about how to proceed.

It was around this time that I found myself thinking about analogies to nuclear weapons, a field in which science advanced in secrecy and without adequate discussions about how researchers' findings should be used. This was particularly true during World War II. J. Robert Oppenheimer, former Berkeley professor of physics and one of the fathers of the atomic bomb, made precisely this point in a series of security hearings following the war, after his outspoken calls for an end to the nuclear arms race (not to mention his Communist ties) had drawn the ire of politicians. Commenting on the American reaction to the Soviet Union's first tests of an atomic bomb and on the ensuing debate over whether to pursue even more explosive hydrogen bombs, Oppenheimer said: "It is my judgment in these things that when you see something that is technically sweet, you go ahead and do it and you argue about what to do about it only after you have had your technical success. That is the way it was with the atomic bomb. I do not think anybody opposed making it; there were some debates about what to do with it after it was made."

Oppenheimer's words only pricked my conscience more. Perhaps one day we would be saying the same thing about CRISPR and genetically modified humans. While human gene editing would almost assuredly never have the same catastrophic consequences as the detonation of a nuclear weapon, it seemed likely that rushing ahead with the research could still cause harm — by undermining society's trust in this new form of biotechnology, if nothing else. Indeed, given the widespread uneasiness about, and even antipathy toward, certain forms of genetic engineering in agriculture, I was becoming especially concerned that a lack of information — or the spread of misinformation — about germline editing could stymie our attempts to use CRISPR in far safer and more essential ways.

As my mind churned through these scenarios, I began to wonder how I could get out in front of the problem. I wanted to find a way to take preemptive action and initiate an honest and open public discourse about this technology I'd helped to create. Could I and other concerned scientists save CRISPR from itself — not after the fact, as had happened with nuclear weapons, but before a cataclysm occurred?

I sought answers in another pivotal moment in the history of biotechnology, an episode when voices of caution had resounded throughout the scientific community and beyond. Then, as now, the cause of concern was a breakthrough in genetic engineering. In this earlier instance, it was the birth of recombinant DNA. And in this case, scientists had moved proactively — and, ultimately, successfully — to prevent their work from inadvertently causing harm.

In the early 1970s, scientists made major advances in the nascent art of gene splicing — chemically fusing, or recombining, purified bits of genetic material from different organisms to create never-before-seen synthetic DNA molecules. Paul Berg, a Stanford biochemist and eventual winner of the Nobel Prize, was the first to achieve this feat, and he did so by combining DNA from three sources: a bacterial virus known as lambda phage, the bacterium *E. coli,* and a monkey virus known as simian virus 40, or SV40. Once he'd combined the viral and bacterial DNA, Berg planned to introduce these hybrid mini-chromosomes into cells so that he could study the functions of individual genes when expressed outside of their normal environment.

But at the time, Berg and other scientists recognized that experimenting with modified genetic material could have myriad, unpredictable, and potentially dangerous consequences. Perhaps most troubling was the thought of what might happen if the synthetic DNA wasn't properly contained and somehow got out of the laboratory. Berg's initial plan had been to transfer the genetic material into lab strains of *E. coli* bacteria, but since the human digestive system naturally harbors billions of innocuous *E. coli,* it seemed plausible that genetically modified *E. coli* might infect and harm humans. Moreover, because the SV40 virus was known to cause tumors in mice, there was a chance that the fragment of SV40 DNA could create a novel carcinogenic pathogen that, if released into the environment, might wreak havoc by spreading cancer-causing genes or antibiotic resistance to humans or some other species.

Because of these concerns, Berg and his team of researchers held off on attempting the experiment. Instead, Berg called for the first of what would eventually become two meetings held in the picturesque Asilomar Conference Grounds, nestled in Pacific Grove, California, on the western tip of the Monterey Peninsula. Before his research went any further, he wanted to enlist his fellow scientists to run a thorough cost-benefit analysis.

The meeting in 1973 — eventually known as Asilomar I — focused on the DNA of cancer viruses and the risks they posed; it did not directly address the new recombinant DNA experiments Berg was considering.

That same year, however, scientists held a second conference focused specifically on gene splicing. The concerns raised at this meeting led scientists to request that the National Academy of Sciences establish a committee to formally investigate the new technology. Berg would serve as the chairman of this group, the Committee on Recombinant DNA Molecules, which met at MIT in 1974. Soon after the meeting, they released a notable report titled "Potential Biohazards of Recombinant DNA Molecules."

The "Berg letter," as it's often called, issued an unprecedented summons for a worldwide moratorium on experiments the committee deemed most hazardous — those aimed at creating antibiotic resistance in new bacterial strains and those aimed at creating DNA hybrids with cancer-causing animal viruses. It was one of the first times that scientists had voluntarily refrained from conducting a whole class of experiments in the absence of any regulatory or governmental sanctions.

The Berg letter also included three other recommendations: first, that scientists adopt a cautious approach to any experiments designed to fuse animal and bacterial DNA; second, that the National Institutes of Health establish an advisory committee to oversee future issues surrounding recombinant DNA; and third, that an international meeting be convened so that scientists from around the world could review recent progress in the field and compare notes on how to deal with potential hazards. This last recommendation would result in the International Congress on Recombinant DNA Molecules, held back in Asilomar in February 1975.

Much has been written about Asilomar II. Roughly a hundred and fifty people attended, mostly scientists but also lawyers, government officials, and members of the media. The debate was heated at times, with even the biology experts disagreeing with one another on the relative hazards of experiments involving recombinant DNA. Some argued against prematurely ending the moratorium, feeling that certain experiments should continue to be prohibited until much more was known about their risks; others felt the risks were likely nonexistent or at least minimal and certainly nothing that the proper safety measures couldn't protect against. Ultimately, Berg and his colleagues decided that most experiments should proceed but with appropriate safeguards; namely, biological and physical barriers to contain genetically modified organisms.

While such resolutions were certainly important, Asilomar II was just as consequential for the link it forged between scientists and the public. The members of the media who attended the meeting informed their audiences about the scientists' discussions. Instead of leading to an uproar and crippling restrictions, as some scientists had feared, this transparency ultimately gave rise to a consensus that allowed research to proceed with popular support.

Asilomar II was not without its critics, though. The conference was invitation-only, and with just a handful of nonscientists in attendance, some argued that the meeting failed to cast a wide enough net outside the scientific community. Others took issue with the omission of topics like biosecurity and ethics from the meeting's agenda. Perhaps the most criticism was reserved for the notion that experts could best assess and address the risks, benefits, and ethical challenges surrounding a new technology, and therefore experts should be the ones to define the terms of the debate. As Benjamin Hurlbut, a science historian at Arizona State University, put it: "This approach gets democracy wrong. It is our technologies that should be subject to democratically articulated imaginations of the futures we want, not the opposite. Science and technology often claim to be servants of society; they should take that promise seriously. Imagining what is right and appropriate for our world — and what threatens its moral foundations — is a task for democracy, not for science."

I absolutely agree that society as a whole — rather than scientists individually or even as a group — should decide how any given technology is used. But there's a wrinkle here, which is that society cannot make decisions about technologies it doesn't understand, and certainly not about those it knows nothing about. It's up to scientists to bring these breakthroughs to the public's attention, as Berg and his colleagues did, to introduce and demystify their technical accomplishments so the public can understand their implications and decide how to use them. When gene splicing was first developed, after all, most biologists weren't even aware of it; the discussion necessarily had to begin within the community of experts who understood what the technology was and what experiments it made possible. By publicizing these discussions and inviting the media to further expound on the technology in terms that laypeople could understand, Berg and colleagues helped break down the wall between scientists and the public and pave the way for the creation of a governmental authority known as the Recombinant DNA Advisory Committee, which became heavily involved in overseeing subsequent research and clinical applications of recombinant DNA.

Some forty years later, in the early part of 2014, I decided that we needed to take a similar approach — not just with CRISPR but with the general practice of gene editing. The technology had already spread like wildfire through the global scientific community; in its brief history, precision gene editing had been used on a diverse and growing menagerie of animals, and all indications were that therapeutic applications in somatic human cells were not far off. But scientists and the public seemed to be ignoring the very real possibility that this same technology would soon be used on human embryos, and they were apparently oblivious to the significance of this sort of germline editing.

An open and frank discussion about germline editing clearly had to begin without delay, and I felt I needed to help initiate the discussion. Much as Berg and his colleagues had sounded the alarm when the risks of their work with recombinant DNA became clear, I would need to leave the comfort of my lab and help spread the word about the implications of our research. Only that way could CRISPR be fully understood by the people whose lives it would soon affect. Only that way, I hoped, could its worst excesses be averted.

It's one thing for a scientist like me to organize an academic meeting on subjects that are firmly within my wheelhouse. It's quite another to take the reins of a conversation about the broader implications of my research, a discussion dealing not with the usual questions of reaction kinetics, biophysical mechanism, and structure-function relationships but with questions of policy, ethics, and regulation. I had never before played that kind of role, and at first I found it extremely intimidating.

Luckily, I didn't need to go at it alone. I'd recently co-founded an institute in the Bay Area called the Innovative Genomics Institute (IGI) with the goal of advancing gene-editing technologies, and I realized that the IGI was perfectly positioned to host a meeting like the ones Berg had hosted at Asilomar. But I knew we'd have to let the conversation evolve organically, not try to go from zero to sixty by holding a lengthy conference right away. I decided we should organize a small, one-day forum and invite around twenty people. The immediate goal, as I saw it, was to produce a white paper — a report proposing a path forward for the field and calling for more stakeholders to weigh in on the issue of gene editing. Much like Berg's 1974 meeting at MIT, this first meeting — which we ultimately called the IGI Forum on Bioethics — would, I hoped, be a prelude to a much larger, more inclusive conference.

We set the meeting date for January 2015 and selected the Carneros Inn in Napa Valley, the renowned wine-growing region just an hour or so north of Berkeley, as the venue. Helping to organize the forum were Jonathan Weissman, a close colleague at the University of California, San Francisco, and codirector of the IGI; Mike Botchan, a Berkeley colleague and IGI administrative director; Jacob Corn, scientific director of the IGI; and Ed Penhoet, professor emeritus at Berkeley and co-founder of the biotechnology firm Chiron. One of the first invitations went to Paul Berg himself (who is professor emeritus at Stanford), and I was thrilled when he accepted. Also on the guest list was David Baltimore, a Nobel Prize–winning biologist at Caltech and a colleague of Berg's; Baltimore had not only attended the MIT meeting in 1974 but also coauthored the resulting paper that called for a moratorium on recombinant DNA research, and he had played a pivotal role in the discussions at Asilomar II. Paul's and David's attendance meant that our meeting would have a direct link to the proceedings that had served as my inspiration. More important, their expertise would undoubtedly help us navigate the difficult terrain ahead.

Also confirmed were Alta Charo, professor of law and bioethics at the University of Wisconsin at Madison; Dana Carroll, one of the gene-editing pioneers in the pre-CRISPR days; George Daley, a stem cell expert from Children's Hospital in Boston; Marsha Fenner, program director of the IGI; Hank Greely, director of the Center for Law and the Biosciences at Stanford University; Steven Martin, professor emeritus and former biological sciences dean at UC Berkeley; Jennifer Puck, professor of pediatrics at UC San Francisco; John Rubin, a film producer and director; Sam Sternberg, my coauthor and PhD student at the time; and Keith Yamamoto, professor at UC San Francisco and administrative director for the IGI. A few other scientists were invited but declined to attend. (George Church and Martin Jinek, two scientists who were not in attendance, ultimately signed the article that was published after the meeting.)

The meeting, which we held on January 24, 2015, featured spirited discussions on a wide range of topics. The attendees, seventeen in all, gave formal presentations on gene therapy and germline enhancement, on existing regulations that governed genetically modified products, and on the nitty-gritty details of CRISPR. Even more interesting than these presentations, in my opinion, were the group's open-table deliberations about the future of gene editing. These conversations were enthusiastic and creative, covering topics I had previously grappled with only on my own.

As we began discussing authorship of a white paper summarizing our conclusions, we debated who our target audience should be and what kind of outcome we were hoping to achieve. Should we be dealing with all the repercussions of using CRISPR — including new kinds of GMOs and even designer organisms — not just its potential role in germline editing? Had CRISPR actually raised new issues about germline modification or were the differences between it and prior technologies only a matter of degree? And would our little group come out strongly against germline editing or would we leave open the possibility for its eventual use?

Over the course of these conversations, a consensus slowly took shape. We decided that the use of gene editing specifically in the human germline should be the focus of our white paper. Gene therapy had been applied to patients' somatic cells for well over two decades, and early gene-editing technologies had also already been used on human somatic cells in clinical trials. It was clear that germline editing was the one area where few had ventured and where public discussion was most urgent. This was largely because CRISPR, we agreed, had lowered the technical barriers that had once made human germline editing difficult, if not impossible, to accomplish. Despite the many volumes previously written on germline modification, and despite the 1998 UCLA conference and the doomsday scenarios explored by science fiction authors over the years, it quite simply hadn't been feasible to edit the human germline with any degree of precision before CRISPR. Now, of course, things were very different — a point driven home by one of the forum's participants, who told us that a scientific manuscript describing experiments in which human embryos were edited with CRISPR was *already* circulating among major journals. This research, if real, would represent the first time that scientists had knowingly tweaked specific DNA sequences in the genome of a potential future human.

If ever there was a time to get the word out, it was now. But what would our position be? Many of us were unsure if it would ever be safe to make heritable changes to the human genome, given that any mistakes could be disastrous for the individual and for future generations. Whether such changes could be ethically justified was another issue entirely. As our conversation stretched into the afternoon, we deliberated over questions of social justice and procreative liberty and openly discussed fears about eugenics. Some participants were wary of science moving in this direction while others admitted that they had no problem with germline editing, at least not in theory. As long as it could be proven safe and effective, this cohort argued, and as long as its benefits clearly outweighed its risks, how could we hold this mode of therapy to a higher standard than any other medical treatment?

Ultimately, though, we realized that this wasn't our decision to make. It was not up to us, the seventeen people in the room, to determine what the public should think about germline editing. We felt that our responsibility was twofold. First, we had to make the public aware that germline editing was an emerging societal issue that should be confronted, studied, discussed, and debated. Second, we had to urge the scientific community — those individuals who *were* familiar with the technology, and who were aggressively pushing it in new directions — to hold off on exploring this one avenue of research. We felt it was critical to discourage our peers from rushing headlong into any research efforts, let alone any clinical applications of gene editing, that involved altering the human germline. Essentially, we wanted the scientific community to hit the pause button until the societal, ethical, and philosophical implications of germline editing could be properly and thoroughly discussed — ideally at a global level.

We pondered how best to achieve these objectives. Should we submit an editorial to a major newspaper? Hold a press conference? Author a perspective — essentially, an academic op-ed — in a scientific journal? After some back-and-forth, we settled on the last option, reasoning that this would likely get the most exposure among active researchers and would probably be picked up by the popular media, as often happened with high-profile articles in major journals. And because our meeting centered on one of the hottest topics in all of biology, we knew this paper would cause a splash.

We concluded the meeting by outlining the paper, which we planned to submit to the journal *Science*. Its goal, we agreed, would be to draw attention to the issue without getting too deep into the weeds. There would, of course, be many highly contentious issues to eventually discuss, but this initial perspective didn't seem like the right place to get into them. We wanted to simply get the ball rolling, and we decided to leave further discussion for a subsequent meeting when more people would be able to attend and participate.

Finally, our energy spent, my colleagues and I adjourned to Angèle, a French restaurant perched above the Napa River. Seated outside around a long oval table, a cool breeze blowing in from the nearby hills, we sipped local wine and snacked on appetizers while enjoying lighthearted conversations about work, family, and travel. We were glad to set aside the heavy topics that had occupied us all morning and afternoon. Yet, privately, my mind was still racing.

Had I really made the right move by entering this new arena? The idea of taking a public stand on a scientific issue, no matter how important, felt foreign to me, almost transgressive. It was unclear whether our perspective would make a lasting impact and whether it would be received as we intended. Even if it went over well, it might be too little, too late. The manuscript our colleague had described, the one that was currently being considered for publication by major science journals, was haunting me. Other such experiments might be in progress at that very moment or be attempted in the near future. Would they be published before we had a chance to announce our conclusions?

I was sure of one thing: now that I had committed to this path, I would move quickly. By the time I made it back home to Berkeley that night, I had already begun organizing my notes and putting together a rough outline. The actual article turned out to be a challenge to write, but within a couple of weeks, I'd sent the first draft to the other Napa forum participants, and we began the round-robin process of editing it. On March 19, 2015, the article was published online with the title "A Prudent Path Forward for Genomic Engineering and Germline Gene Modification."

The article, which ran just a few pages, explained the technology and stated our concerns about it. After introducing CRISPR, the concept of gene editing, and the applications that were currently being pursued, we turned to the topic of germline editing. On that subject, we put forth four specific recommendations. We asked experts from the scientific and bioethics communities to create forums that would allow interested members of the public to access reliable information about new gene-editing technologies, their potential risks and rewards, and their associated ethical, social, and legal implications. We called on researchers to continue testing and developing the CRISPR technology in cultured human cells and in nonhuman animal models so that its safety profile could be better understood in advance of any clinical applications. We called for an international meeting to ensure that all the relevant safety and ethical implications could be openly and transparently discussed — not just among scientists and bioethicists, but also among the many diverse stakeholders who would surely want to weigh in: religious leaders, patient- and disability-rights advocates, social scientists, regulatory and governmental agencies, and other interest groups.

Last, and perhaps most significant, we asked scientists to refrain from attempting to make heritable changes to the human genome. Even in countries with lax regulations, we wanted researchers to hold off until governments and societies around the world had a chance to consider the issue. Although we ultimately avoided using the word *ban* or *moratorium*, the message was clear: for the time being, such clinical applications should be off-limits.

Any fears I'd had about the reception and immediate impact of our article vanished as soon as the perspective was published. In the days that followed, colleagues reached out to thank us for bringing up this issue and to inquire about the meeting to come. Would it be hosted by professional societies or national academies? How did we plan to include countries other than the United States? Would we return to Asilomar for another historic conference or pick another venue? Messages also poured in from journalists and members of the public, thanks in large part to the press our article attracted. The *New York Times* ran a front-page story that generated hundreds of reader comments, and our perspective was also picked up by media outlets from National Public Radio and the *Boston Globe* to numerous blogs and websites. It certainly helped that a team writing in the journal *Nature* had called for a ban on germline editing just days before we did and also that the *MIT Technology Review* had recently published a riveting piece on germline editing.

The topic, it seemed, had suddenly entered the mainstream. In the blink of an eye, CRISPR had morphed from a revolutionary but relatively esoteric technology into a household word. Now that the technology's extraordinary implications for the future of humanity were out in the open, I allowed myself to hope that we could have a broad, frank conversation about germline editing: when, if ever, we would sanction its use, how we would regulate it, and what repercussions we were and were not prepared to tolerate. It was exhilarating to have finally begun the process of public discussions about CRISPR — but the path ahead would be long.

8

WHAT LIES AHEAD

I'D GOTTEN A BAD FEELING in the pit of my stomach when, at the Napa Valley conference, one of my colleagues revealed that CRISPR had already been used to experiment on the genomes of human embryos. Subsequently, I'd heard more rumors about the research and the article describing it, leaving me wondering about the details and even the veracity of the story. What if the rumors were unfounded, fanned by the apprehension of people like me who felt that this sort of research should not proceed unfettered?

As I thought about it, the implications of *any* research involving gene editing of the human germline became more and more troubling. Even if the embryos weren't used to create a living person (which they surely wouldn't be, I told myself, given the massive public backlash this would cause), editing them with CRISPR would still represent a major scientific milestone — the first time the DNA of unborn humans had been subjected to gene editing. Not only would an experiment like this fling open a door we would never be able to close, but it might also scramble the constructive dialogue that my colleagues and I were trying to start. By announcing that their research had already outpaced public debate, the scientists behind these experiments would surely garner a lot of attention and, possibly, spur significant outrage. My biggest worry was that they might inadvertently set many members of the public against this fledgling technology despite its enormous potential for good.

I didn't have to wait long to learn the details of the experiments. On April 18, 2015, just one month after my colleagues and I published our call asking scientists to refrain from clinical use of human germline editing, the rumored scientific article was published. Although the experiments it described were not designed to create embryos that could be implanted in a mother's uterus and carried to term, the study nonetheless attracted substantial attention.

The article, published in the journal *Protein and Cell,* described experiments in Junjiu Huang's lab at Sun Yat-sen University in Guangzhou, China. Huang and his colleagues had injected CRISPR into eighty-six human embryos. The target in this study was the gene responsible for producing beta-globin, a part of the hemoglobin protein that carries oxygen through the body. People with defects in the beta-globin gene develop the debilitating blood disorder known as beta-thalassemia. Huang's goal was to precisely edit the beta-globin gene in these eighty-six embryos, providing proof-of-principle evidence that the disease could be stopped before it ever started.

In an effort to obtain that evidence, Huang's team injected the embryos with humanized genetic instructions to produce the necessary CRISPR molecules — an RNA molecule specifying the proper GPS coordinates within the genome, and the Cas9 protein to slice the gene at that site. Also included was a piece of synthetic DNA with which to repair the broken gene, as well as a jellyfish gene encoding green fluorescent protein. This last ingredient allowed the researchers to analyze the embryos that continued to grow and divide after the gene editing was done; all they had to do was look for glow-in-the-dark cells.

In purely scientific terms, the results of Huang's experiments were mixed. Upon examining the beta-globin genes of the tested embryos, the researchers found that a mere four of the eighty-six embryos contained the intended mutations, a gene-editing efficiency of just 5 percent. There were other problems too. The method turned out to be sloppy. In some embryos, CRISPR edited off-target DNA sequences — that is, the wrong

genes — and peppered the embryos' genomes with unintended mutations. In other embryos, CRISPR correctly sliced apart the intended DNA
sequence in the beta-globin gene, but the cells hadn't healed themselves
properly; instead of using the template supplied by the researchers, they
had repaired the damaged beta-globin gene using sequences from a related gene called delta-globin. On top of these shortcomings, some of the
developing embryos were mosaic, meaning that their cells contained a
hodgepodge of differently edited versions of the beta-globin gene. In one
example, the embryo had at least four different edited DNA sequences,
only one of which was the correct one. Rather than editing the beta-globin gene at the one-cell stage and thus repairing the sole master copy of
the embryo's genome, CRISPR had acted too slowly and begun working
only after the fertilized egg had split into multiple daughter cells.

These were precisely the kinds of safety risks that had motivated me
to publicly call for a halt on experimentation leading to clinical use of
germline gene editing. To be fair, Huang's team recognized that the technique was far from perfect, noting that there was a "pressing need to further improve the fidelity and specificity of the CRISPR/Cas9 platform"
before any clinical applications were attempted. But the fact remained
that we were across the threshold — now that germline editing had been
performed on human embryos in a lab, I figured that it was only a matter
of time until it was tried in a clinical setting.

In this case, at least, Huang had taken care to ensure that no CRISPR
babies could be born as a result of his experiments by using triploid human embryos. So named because they contain three sets of twenty-three
chromosomes (sixty-nine in total) instead of the normal two (forty-six
in total), triploid embryos are nonviable, and in IVF procedures, it's easy
for doctors to identify triploid embryos and discard them well before
implantation.

Huang had seen in these nonviable embryos a perfect model for testing the effectiveness of CRISPR. For the purposes of his experiments, the
triploid embryos were not much different than regular, viable ones. By

using triploid embryos that were destined for disposal, however, Huang and his colleagues could neatly sidestep the inevitable objections that they were destroying potential human lives. And the researchers had gotten explicit consent from patients donating the embryos, had secured approval from an ethics committee, and had fully complied with existing regulations in China. The experiments, I knew, would also have been legal in the United States.

I read the article in my Berkeley office. When I finished, I stared out across the San Francisco Bay, lost in thought. I felt awestruck, and a bit queasy.

Much as I wanted to avoid the thought, it was clear that my work with Emmanuelle, which began with a completely different goal, had led directly to Huang's. What other scientists, what other as-yet-unperformed experiments, would be linked to us?

It quickly became apparent that many others in the scientific community shared my concern about Huang's experiments, even if it didn't strike quite as personal a chord with all of them. I learned that the prestigious journals *Nature* and *Science* had both rejected Huang's manuscript, partly because they had ethical objections to the experiments it described. Many scientists agreed that the research had been pursued prematurely, and others wondered about the motivations behind it. Harvard researcher George Daley told the *New York Times* that the attention scientists were sure to receive for editing the human germline might be "the sort of deranged motivation that sometimes prompts people to do things."

The response to Huang's article from many scientific and governmental agencies was swift, and unanimous. The American Society of Gene and Cell Therapy, the premier professional organization for DNA-based medicine, reaffirmed its support for a "strong stance against gene editing in, or gene modification of, human cells to generate viable human [fertilized eggs] with heritable germ-line modifications." The International Society for Stem Cell Research echoed that sentiment, its president maintaining that "a moratorium on any clinical application of gene editing human embryos is critical." Even President Barack Obama's administration entered the fray. In a blog post titled "A Note on Genome Editing," John Holdren, director of the White House Office of Science and Technology Policy, declared that "the Administration believes that altering the human germline for clinical purposes is a line that should not be crossed at this time." Francis Collins, the director of the National Institutes of Health, took a similar position while further stipulating that the NIH would not provide governmental funding for any research involving the gene editing of human embryos.

American spy agencies seemed rattled by the experiments too. I was shocked when the next Worldwide Threat Assessment — the annual report presented by the U.S. intelligence community to the Senate Armed Services Committee — described genome editing as one of the six weapons of mass destruction and proliferation that nation-states might try to develop, at great risk to America. (The others were Russian cruise missiles, Syrian and Iraqi chemical weapons, and the nuclear programs of Iran, China, and North Korea.) "Biological and chemical materials and technologies, almost always dual use, move easily in the globalized economy," wrote the report's authors, *dual use* being a term of art for technologies that can be used for both peace and war. "The latest discoveries in the life sciences also diffuse rapidly around the globe." Without explaining exactly how genome editing could be weaponized, the report noted that "research in genome editing conducted by countries with different regulatory or ethical standards than those of Western countries probably increases the risk of the creation of potentially harmful biological agents or products. Given the broad distribution, low cost, and accelerated pace of development of this dual-use technology, its deliberate or unintentional misuse might lead to far-reaching economic and national security implications." Lest there be any doubt about the immediate source of their concern, the authors pointedly mentioned that "advances in genome editing in 2015 have compelled groups of high-profile US and European biologists to question unregulated editing of the human germline" — an obvious reference to Huang's article and our perspective.

I was glad to learn that leaders in so many fields shared our sense of urgency about the issue of germline editing, but I was also astounded by warnings like the ones in James Clapper's threat assessment. Pondering possible misuses of CRISPR on my own, I had imagined what rogue scientists might do with it, even had nightmares of Hitler getting his hands on the technology. But what if *living* dictators or terrorists attempted to bend CRISPR to their twisted purposes? How could we possibly stop them? And how would I be able to live with the knowledge that my research, which had stemmed from a desire to understand the natural world and, ultimately, improve human lives, had been used to harm them instead?

I was also struck by the fact that the responses to the first tests of CRISPR in human embryos were far from unanimously negative. In July 2015, writing in the same scientific journal that had published Huang's study a few months earlier, Julian Savulescu, a distinguished philosopher and bioethicist, asserted that there was a moral imperative to aggressively *continue* pursuing similar lines of experimentation. Noting (with considerable oversimplification) that gene editing could "virtually eradicate genetic birth defects" and significantly lower the harm caused by chronic diseases, Savulescu and his coauthors argued that "to intentionally refrain from engaging in life-saving research is to be morally responsible for the foreseeable, avoidable deaths of those who could have benefitted. Research into gene-editing is not an option, it is a moral necessity." A month later, Steven Pinker, the acclaimed Harvard scholar, vented his general frustration at the overly cautious reactions to biotechnological advances like CRISPR in an opinion article in the *Boston Globe*. Instead of creating red tape or introducing prohibitive regulations, he argued that "the primary moral goal for today's bioethics can be summarized in a single sentence. Get out of the way."

Other thought leaders enthusiastically supported gene-editing experiments on embryos but drew a sharp distinction between research and clinical applications. For example, in their statement on genetic modification of the human germline, the Hinxton Group — a global network of ethicists, scientists, lawyers, and policy experts — extolled the tremendous promise of gene editing for human health and recommended that basic research continue unimpeded, using both nonviable and viable embryos. And while they acknowledged that they couldn't yet condone gene editing in the clinic, they noted that "when all safety, efficacy and governance needs are met, there may be morally acceptable uses of this technology in human reproduction, though further substantial discussion and debate will be required."

In short, Huang's paper had opened the floodgates, unleashing a wave of public opinion that had washed away any hopes that we would reach a quick, broad consensus.

Concerned scientists, civic leaders, and members of the public would need to move quickly to initiate the global conversation that my Napa Valley coauthors and I had called for. We had barely published our perspective before Huang's paper came out, and the debate over germline editing was intensifying rapidly. Adding to the furor were rumors that multiple other Chinese groups were already planning or performing their own CRISPR experiments in human embryos. They weren't alone; in September 2015, we learned that scientists at the prestigious Francis Crick Institute in London had asked for regulatory permission to do the same. The field obviously wasn't going to wait for scientists or the public to reach an increasingly elusive consensus.

Luckily, the gears were already in motion for the first international summit on human gene editing. In the late spring and early summer, my fellow organizers and I were ironing out the basic details, including where and when it would be held and under what group's or groups' auspices. Eventually, the U.S. National Academies of Sciences, Engineering, and Medicine agreed to host the summit in Washington, DC, that December — a result that pleased me immensely, since the backing of such distinguished organizations would surely lend credibility to the conference. I was even more excited when both the Chinese Academy of Sciences and the Royal Society (the premier scientific society in the United Kingdom) agreed to join as cohosts. Many of the world's leading gene-editing researchers were based in the United States, the United Kingdom, or China, and the involvement of organizations in each of these countries would send a strong signal to the rest of the world that germline editing was an urgent topic that deserved international discussion — and also that it was far too big an issue to be dealt with piecemeal, by individual countries or organizations.

As we nailed down these details, the other eleven members of the organizing committee and I were also hammering out an agenda for the summit. Our primary goals were to educate the audience about the science of gene editing, discuss the societal implications of this new technological power, and address issues of equity, race, and disability rights. This wide spectrum of topics could be grouped into three basic categories: safety considerations, ethical considerations, and regulatory considerations.

On the safety front, we needed to discuss whether gene editing in the germline could ever be proved safe enough to justify clinical uses. Eventually the potential benefits might outweigh the many possible risks, including those that Huang's study had so clearly revealed. But unintended consequences and how they could be controlled for needed to be considered. In addition, I was unclear if our collective knowledge of human genetics would ever be sufficiently advanced to allow us to anticipate and avoid the worst of those negative effects.

We would also need to grapple with tricky ethical issues, many of which have been hallmarks of divisive debates over other controversial topics, such as abortion, reproductive cloning, and stem cell biology. Was it inherently wrong to experiment on embryos, regardless of whether or not a pregnancy was intended? Could germline editing unfairly predetermine a future child's genetic condition or further marginalize individuals with certain genetic disorders? If abused, could it revive the ugly practice of eugenics that had repeatedly stained scientific history over the past century?

Finally, we needed to talk about legal frameworks to control this powerful new technology. In particular, the roles that governments and scientific communities would play in regulating germline editing had to be examined. Some uses of germline gene editing (for instance, to prevent a child from inheriting a disease) might be considered acceptable, while others (such as genetic enhancement) might be prohibited. And many people including me wondered how important it would be to reach international consensus — and what would happen if we couldn't.

To help us work through these complicated issues, we reached out to experts from a wide range of fields. Invitees included Maria Jasin and Dana Carroll, two pioneers of the use of DNA-cutting enzymes for gene editing; Emmanuelle Charpentier, my CRISPR collaborator; Feng Zhang and George Church, two innovators of gene-editing technologies; Fyodor Urnov, a developer of the first gene-editing drug to reach clinical trials; Daniel Kevles, an expert in the history of eugenics; John Harris, a philosopher and supporter of human enhancement; Marcy Darnovsky, executive director of the Center for Genetics and Society; Catherine Bliss, an expert in gender and sexuality; and Ruha Benjamin, a scholar of race-ethnicity, health, and biotechnology. Representing government and legal interests were Congressman Bill Foster (D-IL); White House science adviser John Holdren; legal experts Alta Charo, Pilar Ossorio, Barbara Evans, and Hank Greely; and representatives from China, France, Germany, India, Israel, South Africa, South Korea, and other countries around the world.

Like the Napa Valley conference, the International Summit on Human Gene Editing aimed to broaden the conversation about germline editing, not conclude it. Indeed, by the end of the meeting, which took place in the first few days of December 2015, I found that I had as many questions as I had had when we'd started, if not more. But I also had a deeper appreciation for the rationale of people on opposing sides of the debate, many of whom argued passionately for their points of view and helped me to refine my own thinking on the issue of germline editing.

It would be impossible to fit all of these conversations and viewpoints into a single book, let alone a single chapter, so I will confine myself to one perspective: my own. In the pages ahead, I'll explain how my opinions have changed as a result of these discussions and as a result of the research and reflection I've been engaged in since the Washington summit. I've been doing my best to sort through the differing takes on the issue and to weigh the pros and cons of each. And while I can't claim to have all the answers, my meditations have led me to some conclusions about how CRISPR might one day be used to edit the genomes of unborn humans, safely and ethically, and where the greatest dangers of germline editing actually lie. I've also had to confront some cold, hard facts about public policy — both its shortfalls at present and the lengths to which we as a civil society must go to make CRISPR the tool for good I so firmly believe it can be. I hope these reflections will help advance the conversation about germline editing — and help us to decide whether, and how, we will intervene in the evolutionary journey of our species.

It's almost certain that germline editing will eventually be safe enough to use in the clinic. Microsurgery on egg cells and embryos, such as fertilization by sperm injection and biopsy sample removal for preimplantation genetic diagnosis (PGD), has already become routine in fertility clinics. Delivery of CRISPR has also been optimized in animal embryos and many kinds of human cells. Perhaps the biggest hurdle is ensuring the accuracy of CRISPR itself. But based on the latest research, it seems like even that challenge — how to make this gene-editing system precise enough to alter only the target gene and only *exactly* as intended — will soon be overcome.

How accurate must CRISPR be in order to be safely used in the human germline? It seems obvious that we should reject any procedure that might trigger DNA editing at unintended sites, as sometimes occurs with CRISPR and other gene-editing technologies. But the truth is that our entire lives are spent at risk of such random genetic changes, and the threat from them is arguably far greater than any that CRISPR would pose.

Our DNA is constantly changing, roiled by random, naturally occurring mutations. These natural mutations are the very drivers of evolution, but they are also how inheritable genetic diseases can arise. Every time our cells duplicate their DNA during cell division, somewhere between two and ten novel DNA mutations creep into the genome. Every person experiences roughly one million mutations throughout the body *per second,* and in a rapidly proliferating organ like the intestinal epithelium, nearly every single letter of the genome will have been mutated at least once in at least one cell by the time an individual turns sixty. This mutational process begins from the earliest moment of fertilization, and as the single-cell zygote goes on to divide into two cells, then four, then eight cells of the growing embryo, the new mutations it has acquired will be faithfully copied into the genome of every descendant cell in that growing person's body. Even the sex cells that create the embryo — the mother's egg and the father's sperm — have incorporated new mutations that never before existed in either family's germline. As a result, each one of us begins life with fifty to a hundred random mutations that arose de novo ("anew") in our parents' germ cells.

Any mutations that CRISPR might make — intentional or not — would almost certainly pale in comparison to the genetic storm that rages inside each of us from birth to death. As one writer put it, "Genetic editing would be a droplet in the maelstrom of naturally churning genomes." If CRISPR could eliminate a disease-causing mutation in the embryo with high certainty and only a slight risk of introducing a second off-target mutation elsewhere, the potential payoffs might well outweigh the dangers.

Even more reassuringly, we have tools to safeguard against these off-target effects — at least, we do where germline editing is concerned.

One such tool is PGD, which could make it possible to detect rare, undesirable mutations after CRISPR has edited the genome but before the growing embryo is placed in the mother's womb. Another option that might become possible in the future is to avoid off-target mutations entirely by editing primordial egg and sperm cells instead of fertilized embryos. Although the technology is still in its infancy, research in mice has demonstrated that eggs and sperm can be grown in the laboratory from stem cells and used to establish pregnancies. By eliminating disease-causing mutations with CRISPR and exhaustively screening for off-target mutations before the moment of fertilization, scientists could ensure that only those sex cells with the desired genome are used for reproduction. While we don't yet have the means to perform that procedure in humans, it seems likely that research advances over the next decade will put it within our reach.

When it comes to assessing the accuracy of germline editing, there are *many* issues to consider, but bleeding-edge science suggests that few, if any, of these issues are likely to be deal-breakers. Given the speed with which scientists have already perfected the procedure in mice and monkeys, and given how close we are to clearing the remaining technical hurdles, it seems undeniable that germline editing, in one form or another, will become reliable enough to perform on humans, or at least will have no more risks than natural reproduction does.

Of course, if we're going to propose making heritable changes in the human germline, then we must consider not just whether the technology can work accurately, but also whether the effects of accurate edits will be the ones we intend. We already know that some of the gene edits scientists are considering for clinical use have secondary effects. For instance, editing an embryo's *CCR5* gene might make the resulting human resistant to HIV but more susceptible to the West Nile virus. Correcting the two mutated copies of the beta-globin gene in people who suffer from sickle cell disease would rid them of the illness but also deprive them of the mutation's protection against malaria. These are far from the only gene edits that have both positive and negative effects. Researchers now suspect that people who carry one copy of the mutated gene that causes cystic fibrosis (which requires two copies) have an increased defense against tuberculosis, an infectious disease that accounted for 20 percent of all European deaths between 1600 and 1900. Even gene variants implicated in neurodegenerative diseases like Alzheimer's may have benefits, such as improved cognitive function and better episodic and working memory in young adults.

The fact is that editing a particular gene will always carry the risk of unforeseen effects. But just because we don't know what that collateral damage might be doesn't mean we should renounce germline editing altogether. As George Church, the celebrated Harvard geneticist, wrote, "The notion that we need complete knowledge of the whole human genome to conduct clinical trials of heritable gene editing seems at odds with medical reality." During the roughly four centuries that it took to eradicate smallpox, he pointed out, we knew precious little about the human immune system. What's more, he observed, in situations where we're attempting to correct harmful mutations, "each edit changes the DNA from a sick version to a healthy version that billions of people share. This is far higher certainty than we have for a completely new drug that has never before been tested on humans."

These points seem incontestable. Countless lifesaving medical treatments were developed well before physicians completely understood them, so why would we hold CRISPR to a higher standard of safety? And as long as we are correcting genetic mutations by restoring the "normal" version of the gene — not inventing some wholly new enhancement not seen in the average human genome — we're likely to be on the safe side. If a person's life is hanging in the balance, the potential payoffs of these sorts of limited procedures may be worth the risks.

If we could judge germline editing by its safety alone, I would be cautiously in favor — but that's far from the only criterion to consider. The prospect of editing an unborn person's DNA forces us to confront all sorts of ethical problems, some of which, when I first glimpsed them, were vexing enough for me to call for a temporary halt to germline gene editing so we could examine them more closely.

Just because we *can* edit the human germline, does that mean that we *should*? It's a question I've asked myself again and again. If CRISPR can indeed help certain parents conceive a disease-free child when no other options exist, and if it can do so safely, ought we to use it?

There are some rare situations in which germline editing would be the only way to guarantee that children would be born without genetic disease. For example, in cases where both parents suffer from the same recessive genetic disorder — conditions like cystic fibrosis, sickle cell disease, albinism, and Fanconi anemia — every single child they conceived through natural reproduction would be fated to have the disease as well. Since the causative gene mutation is present on both copies of both parents' chromosomes, the child would have no way to avoid inheriting two mutated copies. A similar scenario presents itself for dominant genetic disorders — conditions like Huntington's disease, the familial form of early-onset Alzheimer's disease, and Marfan syndrome — in which a single copy of the mutated gene is sufficient to cause disease, regardless of whether it comes from the father or mother.

Although these diseases could still be treated with therapeutic gene editing in somatic cells, germline editing would prevent children from developing the diseases in the first place and thus could prevent suffering. In such scenarios, germline editing would indeed seem justifiable from a medical-need perspective — but as I said, they are rare. Much more common are cases where genetic disease is a risk, but not a certainty. In these situations, would germline editing also be justifiable? And if we consider both types of scenarios together, on balance, would germline editing be good or bad? Would it alleviate more suffering than it caused?

That question — "Should we or shouldn't we?" — has gripped scientists and laypeople alike. Perhaps unsurprisingly, America is having a hard time agreeing on an answer; a 2016 Pew Research poll found that 50 percent of adults in the U.S. oppose the idea of reducing the risk of disease using germline editing, compared to 48 percent in favor. (When it comes to making nonessential enhancements to a baby's genome, we seem to be considerably more unified; only 15 percent of the poll's respondents were in favor of that.) Various considerations are guiding these responses.

Religion is one obvious moral compass that people use when confronting difficult questions like this, though perspectives can vary widely. When it comes to experimentation with human embryos, many Christian communities are opposed because they regard the embryo as a person from conception, whereas Jewish and Muslim traditions are more accepting because they do not consider embryos created in vitro to be people. And while some religions see any interventions in the germline as a usurpation of God's role in humanity's existence, others welcome human involvement in the works of nature as long as the goals being pursued are inherently good, such as improved health or fertility.

Another moral guidepost is purely internal: the visceral, knee-jerk reaction to the idea of using CRISPR to permanently edit a future child's genes. For many people, the very idea feels unnatural and somehow wrong, and when I first started thinking about human germline editing, I was one of these people. Humans have been reproducing for millennia aided only by the DNA mutations that arise naturally, and for us to begin directing that process rationally — similar to the way plant biologists might genetically modify corn — seemed almost perverse at first glance. As NIH director Francis Collins put it, "Evolution has been working toward optimizing the human genome for 3.85 billion years. Do we really think that some small group of human genome tinkerers could do better without all sorts of unintended consequences?"

While I share the general feeling of unease at the idea of humans taking control of their own evolution, I wouldn't go so far as to say that nature has somehow fine-tuned our genetic composition. Obviously, evolution didn't optimize the human genome for the present era, when modern foods, computers, and high-speed transportation have completely transformed the way we live. And if we look over our shoulders at the course of evolution that has led to this moment, we'll see that it's littered with organisms that certainly didn't benefit from the mutational chaos that underpins evolution. It turns out nature is less an engineer than a tinkerer, and a fairly sloppy one at that. Its carelessness can seem like outright cruelty for those people unlucky enough to inherit genetic mutations that turned out to be suboptimal.

Similarly, the argument that germline editing is somehow unnatural doesn't carry much weight with me anymore. When it comes to human affairs, and especially the world of medicine, the line between natural and unnatural blurs to the point of disappearing. We wouldn't call a coral reef unnatural, but we might use the term for a megalopolis like Tokyo. Is this because one is crafted by humans and the other isn't? In my mind the distinction between natural and unnatural is a false dichotomy, and if it prevents us from alleviating human suffering, it's also a dangerous one.

I've now had numerous opportunities to meet with people who have experienced genetic disease themselves or in their families, and their stories are deeply moving. One woman pulled me aside at a conference, after a session in which I had discussed CRISPR technology, to share her personal story. Her sister had suffered from a rare but devastating genetic disease that affected her physical and mental health and caused tremendous hardship for the entire family. "If I could use germline editing to remove this mutation from the human population so that no one else suffers as my sister did, I would do it in a heartbeat!" she said, tears welling up in her eyes. On another occasion, a man came to visit me at Berkeley and explained that his father and grandfather had died of Huntington's disease and that three of his sisters had tested positive for the trait. He wanted to do anything he could to advance research toward a cure or, better yet, prevention of this terrible disease. I did not have the heart to ask him if he too carried the mutated gene. If he did, he could expect to be robbed of his powers of movement and speech before much longer and to meet an early death — a terrible sentence for anyone to see levied on their loved ones, let alone be subjected to themselves.

Stories like these underscore the terrible human costs of genetic diseases — and of hesitating to confront them. If we have tools that can one day help doctors safely and effectively correct mutations, whether prior to or just after conception, it seems to me that we'd be justified in using them.

Not everyone shares these views. It's not uncommon to hear people talk about our genomes as if they were part of a precious evolutionary inheritance, something to be cherished and conserved. For example, in its Universal Declaration on the Human Genome and Human Rights, adopted in 1997, the United Nations Educational, Scientific and Cultural Organization (UNESCO) wrote that "the human genome underlies the fundamental unity of all members of the human family, as well as the recognition of their inherent dignity and diversity. In a symbolic sense, it is the heritage of humanity." In light of recent advances in gene editing, UNESCO has further argued that, while technologies like CRISPR should be used to prevent or treat life-threatening diseases, implementing them in a way that would affect future descendants would "jeopardize the inherent and therefore equal dignity of all human beings and renew eugenics, disguised as the fulfilment of the wish for a better, *improved* life." Some bioethicists have voiced similar concerns, suggesting that germline editing changes the very nature of what it means to be human and that modifying the human gene pool would perniciously alter humanity itself.

Philosophical objections like these are worth contemplating. But when I think about the pain that genetic diseases cause families, the stakes are simply too high to exclude the possibility of eventually using germline editing.

Setting aside the inherent rightness or wrongness of editing the germline, two other ethical issues continue to nag at me. Both were discussed at the international summit on gene editing; neither has been resolved. The first has to do with whether anyone will be able to control how germline editing is employed once doctors start using it to save people's lives. The second has to do with questions of social justice — of how CRISPR would affect society.

First, if we agree to use CRISPR in the germline to eliminate genetic diseases, we have to acknowledge that it might also be used to create genetic enhancements — changes in which DNA is altered not to correct a harmful gene variant but to provide some type of genetic advantage.

Of course, there is a limit to what enhancements will be possible or safe to attempt. Many kinds of enhancements that come to mind — things like high intelligence, prodigious musical ability, mathematical prowess, tall stature, athletic skill, or stunning beauty — don't have clear-cut genetic causes. That's not to say they aren't heritable, just that the complexity of these traits may place them beyond the reach of a tool like CRISPR.

But plenty of other genetic enhancements do result from simple mutations that could be re-created with CRISPR. For example, mutations in the *EPOR* gene, which responds to the hormone erythropoietin (the famous doping drug used by Lance Armstrong and countless other athletes), confer exceptional levels of endurance; mutations in a gene called *LRP5* endow individuals with extra-strong bones; mutations in the *MSTN* gene (the same *myostatin* gene that's been edited to create supermuscular pigs and dogs) are known to result in leaner muscles and greater muscle mass; mutations in a gene called *ABCC11* are associated with lower levels of armpit odor production (and, oddly, the type of earwax an individual produces); and mutations in a gene called *DEC2* are associated with a lower requirement of daily sleep.

Ironically, allowing germline editing in cases where it prevents disease might be the first step down a slippery slope to blatantly nonmedical enhancements. That's because for every straightforward example of a non-medical genetic enhancement, there's another that's more ambiguous.

One of these liminal examples of germline editing involves the gene *PCSK9*, which produces a protein that regulates a person's level of low-density lipoprotein cholesterol (the "bad" cholesterol), making the gene one of the most promising pharmaceutical targets to prevent heart disease — the leading cause of death worldwide. CRISPR could be programmed to tweak this gene and save unborn people from high cholesterol. Would this qualify as therapeutic germline editing or enhancement gene editing? Ultimately, the intended goal would be to prevent disease, but it would also endow a child with an advantageous genetic trait that most others don't have.

Plenty of other potential applications for germline editing blur the line between therapy and enhancement. Editing *CCR5* with CRISPR could confer lifelong resistance to HIV; editing the *APOE* gene could lower an individual's risk of developing Alzheimer's disease; altered DNA sequences in *IFIH1* and *SLC30A8* could lower a person's risk of developing type 1 and type 2 diabetes; and changes to the *GHR* gene could reduce an individual's risk of cancer. In all these cases, the primary objective would be to save an individual from disease, but scientists would accomplish this by providing intrinsic protections above and beyond the average person's genetic endowment.

This brings me to my other concern: just as it's hard to know where we'd draw the line when it comes to editing embryos, it's difficult to see how we'd do it equitably — that is, in a way that improves human health across the board, not just in certain groups.

It's not a stretch to think that wealthy families would benefit from germline editing more than others, at least in the beginning. Recent gene therapies that have hit the market come with a price tag of around a million dollars, and it's likely that the first gene-editing therapies will be no different.

Of course, new technologies shouldn't be rejected simply because they are expensive. You need look no further than personal computers, cell phones, and direct-to-consumer DNA sequencing to see how costs of new technologies generally diminish over time as improvements are made, leading to a resulting increase in access. Furthermore, there's also the chance that germline editing, like other medical treatments, could one day be subsidized by health insurance. This certainly seems like only a remote possibility in the United States, since existing reproductive procedures such as IVF and PGD, which routinely cost tens of thousands of dollars, are seldom covered by health insurance. But in places like France, Israel, and Sweden, countries whose national health plans cover assisted reproduction, it's possible that simple economics will incentivize governments to make gene editing available to patients who need it. After all, providing lifelong treatment to a single person with a genetic disease could be much more expensive than prophylactic intervention in the embryo using gene editing.

But even in countries with comprehensive health-care systems, where people from all classes could benefit from germline editing, there's a risk that it might give rise to hitherto unseen genetic inequalities, creating a new "gene gap" that would only widen over time. Since the wealthy would be able to afford the procedure more often, and since any beneficial genetic modifications made to an embryo would be transmitted to all of that person's offspring, linkages between class and genetics would ineluctably grow from one generation to the next, no matter how small the disparity in access might be. Consider the effect this could have on the socioeconomic fabric of society. If you think our world is unequal now, just imagine it stratified along both socioeconomic *and* genetic lines. Envision a future where people with more money live healthier and longer lives thanks to their privileged sets of genes. It's the stuff of science fiction, but if germline editing becomes routine, this fiction could become reality.

Germline editing may inadvertently transcribe our societies' financial inequality into our genetic code — but it may also create a different kind of injustice. As disability-rights advocates have pointed out, using gene editing to "fix" things like deafness or obesity could create a less inclusive society, one that pressures everyone to be the same — and perhaps even encourages more discrimination against differently abled people — instead of celebrating our natural differences. After all, the human genome is not mere software with bugs that we should categorically eliminate. Part of what makes our species unique, and our society so strong, is its diversity. While some disease-causing gene mutations produce defective or abnormal proteins on a biochemical level, the individuals that live with the disease are certainly not defective or abnormal, and they might live perfectly happy lives and not feel any need for gene repair.

This fear — that gene editing will exacerbate existing prejudices against people who fall outside a narrow range of genetic norms — underlies the association that numerous writers have made between germline editing and eugenics. That concept is best known today for its popularity in Nazi Germany, where a quest to perfect the human race reached its terrible zenith through the forced sterilization of hundreds of thousands of people and the widespread extermination of millions of Jews, homosexuals, the mentally ill, and others deemed unworthy of life. Sadly, similar eugenics practices were already common in the United States well before Hitler came to power, and compulsory sterilization continued in numerous states through the 1970s. Given the deplorable history humans have when it comes to programs aimed at improving our species' gene pool, perhaps it's not surprising that CRISPR's potential to endow individuals with healthier genes earns it comparisons to these sad chapters in our past.

But while it's certainly attention-grabbing to equate gene editing with these dark precedents, the comparison doesn't hold up to scrutiny. Technically, the use of CRISPR in embryos to combat human disease *would* be a eugenic practice, but so is preimplantation genetic diagnosis, ultrasound technology, prenatal vitamins, and even a mother's abstinence from alcohol during pregnancy. That's because *eugenic*, as it was originally defined, means "well-born" and so could apply to any action intended to lead to the birth of a healthy child. Our current, looser interpretation of the term reflects beliefs and practices that emerged in the late nineteenth century and the first half of the twentieth century and that were aimed at improving the genetic qualities of the entire population by encouraging breeding between people with desirable characteristics and discouraging or preventing reproduction by people who were deemed undesirable.

Eugenics, as most people remember it today, was certainly reprehensible, but the odds are minuscule that we'll see anything similar happen with gene editing. Governments are simply not going to begin forcing parents to edit their children's genes. (In fact, as we'll see in the next section, the procedure is still illegal in many places.) Unless we're talking about coercive governments controlling their citizens' procreative liberty, germline editing would remain a private decision for individual parents to make for their own children, not a decision for bureaucrats to make for the population at large.

My views on the ethics of germline editing continue to evolve — but as they do, I find myself returning time and again to the issue of choice. Above all else, we must respect people's freedom to choose their own genetic destiny and strive for healthier, happier lives. If people are given this freedom of choice, they will do with it what they personally think is right — whatever that may be. As Charles Sabine, a victim of Huntington's disease, put it, "Anyone who has to actually face the reality of one of these diseases is not going to have a remote compunction about thinking that there is any moral issue at all." Who are we to tell him otherwise?

I don't believe there's an ethical defense for banning germline modification outright, nor do I think we can justifiably prevent parents from using CRISPR to improve their chances of having a healthy, genetically related child, so long as the methods are safe and are offered in an equitable manner. I also don't see how we can allow germline editing to proceed unless we make a conscious effort to support parents' choice to procreate the old-fashioned way and unless we redouble our commitment to building a society in which all humans are respected and treated equally, regardless of their genetic makeup. If we can do these things — if we can walk the narrow line between prohibiting CRISPR to the detriment of certain individuals' health and overusing it and subverting our society's values — we will be able to use this new technology in a way that is unequivocally good.

So how do we ensure that happens? Starting conversations about ethics and safety is one thing; coming to an agreement is quite another. Going a step further and actually acting on a resolution, if we can even get to one, might seem so remote a possibility as to hardly be worth talking about. But if we don't start planning for consistent, international guidelines now, we might not get a second chance.

There can be no doubt that governments have a role to play in overseeing and regulating methods that alter the human germline. But there's a lot of work to be done here, since current government regulations on the topic are variable and often lack teeth. For example, in a long list of countries that includes Canada, France, Germany, Brazil, and Australia, clinical interventions in the human germline are expressly prohibited, with criminal sanctions that range from fines to lengthy prison terms. In other countries, such as China, India, and Japan, these interventions are forbidden, but with guidelines that are not legislative and thus less enforceable. In the United States, the current policy might be considered restrictive: there are no outright bans, but government agencies have spoken out against clinical use of gene editing in the germline, and any clinical trials would need to receive regulatory approval by the Food and Drug Administration. (It's interesting to note, though, that many other assisted reproductive technologies — preimplantation genetic diagnosis, intracytoplasmic sperm injection, and even the practice of in vitro fertilization itself — never underwent formal clinical trials or FDA review.)

Even regulations about *research* into human germline editing — much less the rules governing the creation of new humans from edited embryos — are inconsistent. In China, where the first experiments using CRISPR in embryos took place, this sort of research may proceed with proper oversight by institutional review boards. The same research is technically unrestricted by the U.S. federal government (though some states do forbid it), but a bill passed in the United States in 1996, the Dickey-Wicker Amendment, prevents the government from funding any research that would create or destroy human embryos, an exclusion that would clearly apply to experiments with CRISPR. No laws in the United States prohibit privately funded research in this area. Germline-editing research is allowed (and is already being conducted) in the United Kingdom, but it requires approval by an organization known as the Human Fertilisation and Embryology Authority. Finally, some governments restrict any research involving human embryos or have vague laws that leave considerable uncertainty over the distinction between clinical and research applications.

The nebulous language in which government policies toward germline editing are often couched makes the issue of regulation particularly challenging. For example, a recently adopted document regulating clinical trials in the European Union prohibits "gene therapy clinical trials . . . which result in modifications to the subject's germline genetic identity." How "genetic identity" is defined is unclear, however, as is the question of whether "gene therapy" encompasses gene editing with CRISPR. In France, acts that "undermine the integrity of the human species" are banned, as is any "eugenic practice aimed at organizing the selection of persons." Yet preimplantation genetic diagnosis — a procedure that falls under the literal definition of *eugenic* — is offered in French clinics, so this is too vague to be useful. In Mexico, by contrast, existing regulations over human genetic manipulation are measured by their purpose; aims other than "the elimination or reduction of serious diseases or defects" are prohibited. But who decides what constitutes a serious disease or defect? The government? Physicians? The parents?

The U.S. Congress has so far been unwilling or unable to even *look* at petitions to use CRISPR clinically in human embryos — a legislative approach that is tantamount to our elected leaders burying their heads in the sand. In 2015, the U.S. House and Senate appropriation bill included a rider that blocked the FDA from using congressional funds to review any application of a drug or biological product "in which a human embryo is intentionally created or modified to include a heritable genetic modification." In other words, members of Congress effectively banned the use of CRISPR in embryos, but they did it by tying the FDA's hands behind its back instead of enacting actual legislation. (In an ironic twist, this roundabout strategy almost backfired, since within the current regulatory process, applications to investigate new drugs are automatically approved in thirty days unless the FDA explicitly rejects them. The workaround for this problem was a last-minute addition in 2016 that instructed the agency to treat such applications as if they had never been received.)

Refusing to even review research about germline editing hardly seems like the best way to regulate the practice. Simply banning research or clinical uses of germline editing isn't the way to go either. As other writers have pointed out, any prohibitions on germline gene editing in the United States would effectively cede leadership in this area to other nations — something Americans are arguably already doing with our existing bans on federal funding for germline editing.

There's also a risk that overly restrictive policies in some countries will encourage what might be called CRISPR tourism in others. Patients with means could travel overseas to jurisdictions where regulations are more forgiving or absent altogether. Medical tourists have already spent millions of dollars to receive unregulated stem cell treatments internationally, and gene therapy treatments to increase muscle mass and lengthen lifespan have also been pursued abroad. The solution to these dangerous and even unethical practices isn't for sovereign nations to simply permit risky, unproven methods to be performed on their soil. Nor is it to impose excessive restrictions on research, since that might lead scientists to continue their experiments behind closed doors — arguably one of the worst possible outcomes. Rather, nations need to maintain regulatory environments that are hospitable enough to permit research and clinical applications but strict enough to prevent the worst excesses.

It's up to researchers and lawmakers alike to find the right balance between regulation and freedom. Scientific experts should work to create a set of standardized, agreed-upon guidelines that specify the safest methods of CRISPR delivery, prioritize disease-causing genes for research, and set quality-control standards to evaluate gene-editing interventions. And government officials — especially in the United States — need to take a more active role than they have so far, pursuing robust legislation while also soliciting the opinions of their constituents and encouraging public participation, much as my colleagues and I tried to do with the 2015 summit in Washington, DC. It's unrealistic to think that there will ever be unanimous agreement on whether and how to use germline editing, of course, but governments should nevertheless do their best to pass laws that both harness its potential and represent the will of the people.

Even with these efforts, it's unlikely we'll see anything close to a coherent international response to the challenges posed by CRISPR. Different societies will inevitably approach the topic of germline editing with unique perspectives, histories, and cultural values. Some authors have predicted that human germline editing, especially genetic enhancement, will be adopted first in Asian countries like China, Japan, and India. China is particularly fertile ground for germline-editing research and development, as its scientists have led the way in employing CRISPR technology in several areas, including the first uses in nonhuman primates, nonviable human embryos, and human patients.

But as elusive as an international accord on germline editing may be, we must try for it. Because while the risk that gene editing will fragment human society seems like a problem for future generations, it looks less remote in light of history.

Once a game-changing technology is unleashed on the world, it is impossible to contain it. Blindly rushing ahead with new technologies creates problems of its own. For example, the race for supremacy in nuclear technology led to massive research-and-development efforts that fundamentally reshaped the global political system and many aspects of people's lives, often in ways that made us less safe. Unlike nuclear technology, gene-editing technology gives us the chance to hold an informed public discussion about how we want to use CRISPR's most far-reaching power: the ability to control the future of life. But if we wait too long, we may find that the reins have slipped from our hands.

One of the defining characteristics of our species is its drive to discover, to constantly push the boundaries of what is known and what is possible. Advances in rocket science and space travel are allowing us to explore other planets, and advances in particle physics are revealing the fundamental underpinnings of matter; in the same way, advances in gene editing are enabling us to rewrite the very language of life — and putting us closer to gaining near-complete control of our genetic destiny. Together, we can choose how best to harness this technology. There's simply no way to unlearn this new knowledge, so we must embrace it. But we must do so cautiously, and with the utmost respect for the unimaginable power it grants us.

For most of our species' history, humans have been subjected to slow, often imperceptible evolutionary pressures exerted by the natural world. Now we find ourselves in the position of controlling the focus and intensity of those pressures. From here, things will progress much more quickly than either our species or our planet is accustomed to. It's hard to predict what the average human genome will look like just a few decades from now. Who's to say how our species or our world will appear in a few hundred years — or a few thousand?

Aldous Huxley famously imagined a future of genetic castes in his chilling novel *Brave New World,* and rarely does the topic of germline gene editing come up in the media nowadays without the book being directly or indirectly referenced. But Huxley's dystopia is set in the year 2540. It seems unlikely that genetic inequality — if it does result from germline editing — will take nearly that long to set in. And just think of all the other ways a technology like CRISPR could redefine our society, and our species, over the span of half a millennium. It's a sobering exercise, to say the least.

Many of these changes will be unequivocally good. CRISPR has such incredible potential to improve our world. Imagine using gene editing to eradicate the most severe genetic diseases, much as vaccination ended smallpox and may soon end polio. Imagine thousands of scientists using CRISPR to study scourges like cancer and coming up with new treatments, even cures. And imagine farmers, breeders, and world leaders solving the world's hunger crisis using CRISPR-generated crops that can better weather our changing climate. These scenarios could be within reach, or not, depending on the choices we make in the years ahead.

Few technologies are inherently good or bad; what matters is how we use them. And when it comes to CRISPR, the possibilities of this new technology — good and bad — are limited only by our imaginations. I firmly believe we can use it for the former and not the latter, but I am also cognizant that this will require determination from us, individually and collectively. As a species, we have never done anything like this before — but then again, we have never had the tools to do it.

The power to control our species' genetic future is awesome and terrifying. Deciding how to handle it may be the biggest challenge we have ever faced. I hope — I believe — that we are up to the task.

EPILOGUE:
THE BEGINNING

AS I WRITE THIS, I am traveling home from New York's Cold Spring Harbor Laboratory, where I attended the laboratory's second annual conference on the gene-editing revolution brought about by CRISPR. The book of abstracts from the conference, containing the summaries of all the scientific research that was presented, is open on my laptop, along with notes from the many discussions I had with meeting participants. The attendees — some 415 in all — included not only researchers from academic and corporate labs, but also medical doctors, journalists, editors, investors, and people afflicted with genetic disorders. I've seen a similar mix of people at the many seminars I've presented at universities and foundations over the past few years. These groups are a cross-section of the stakeholders who will be affected by gene-editing technologies and who will help shape their future uses.

While I was at Cold Spring Harbor, a student — visibly pregnant — introduced herself to me and asked if I could summarize my personal journey as a scientist and mother living in the midst of the CRISPR revolution. Thinking about the metaphorical distance I had traveled, I had to laugh. And then I gave it a try.

It's been a roller-coaster ride, one whose twists and turns I could not have imagined at the outset. I've experienced the pure joy of discovery; the "pleasure of finding things out," as physicist Richard Feynman put it. I have marveled with my son about the ways that bacteria program proteins to work like armed guards that recognize and destroy invading viruses. I have relished being a student again, learning about topics like human development and the medical, social, political, and ethical issues that attend human reproduction. I have rediscovered what a special person I am married to, a partner who is wise, supportive, and deftly capable of everything from running a world-class research laboratory to helping our son with his latest rocket-building attempt to interpreting legal documents submitted to the U.S. Patent and Trademark Office. He also cooks a mean mushroom quesadilla and has excellent taste in Chianti.

Over the past four years (and, indeed, throughout my career), I have had the privilege and honor of working with some of the best and most brilliant scientists in the world. In my own lab, I was incredibly lucky to benefit from the hard work and dedication of countless students, postdocs, and staff scientists, folks like Blake Wiedenheft, Rachel Haurwitz, Martin Jinek, and my coauthor, Sam Sternberg, who were the ones actually running experiments on a day-to-day basis. Outside of my lab, I delighted in the opportunity to work with luminaries of the science world like Paul Berg and David Baltimore, who helped guide our quest to initiate a public conversation about the implications of gene editing, and fantastic collaborators like Jill Banfield and Emmanuelle Charpentier, who challenged me to pursue new avenues of research.

Of course, while collaborations grease the wheels of scientific research, competition is often the fire that stokes the engine. Healthy rivalries are a natural part of the scientific process, and they have fueled many of humankind's greatest discoveries. But at times, I have been taken aback by just how intensely competitive the study and use of CRISPR can be and how much it transformed in a matter of years, becoming a global field that touches virtually any researcher studying biology.

These twin poles of science — competition and collaboration — have defined my career and shaped me as a person. Over the past half decade in particular, I have experienced the gamut of human relationships, from deep friendships to disturbing betrayals. These encounters taught me about myself and showed me that humans must choose whether they will control or be controlled by their own aspirations.

I have also come to appreciate the importance of stepping out of my comfort zone and discussing science with people beyond my circle of specialists. Scientists are viewed with increasing distrust by a public that is skeptical about their contributions to society — that is, skeptical about the power of science to describe and improve the world. When people refuse to acknowledge climate change, reject vaccination programs for children, or insist that genetically modified organisms are unfit for human consumption, it signals not only their ignorance about science, but also a breakdown in communication between scientists and the public. The same can be said of the protest movements against CRISPR that have already sprung up in France and Switzerland to decry the prospect of "GM babies." Unless we can reach these people and others like them, such distrust will spread.

Scientists are partially responsible for this breakdown in communication. I had difficulty wrenching myself out of the lab to talk about the implications of CRISPR, and sometimes I wish I had done it sooner. I've come to feel strongly that we who practice science are obligated to participate actively in discussions about its uses. We live in a world where science is global, where materials and reagents are distributed by central suppliers, and where it is easier than ever to access published data. We need to make sure that knowledge flows just as freely between scientists and the public as it does among the researchers themselves.

Given how radical the implications of gene editing are for our species and our planet, opening the lines of communication between science and the public has never been more essential than it is now. Gone are the days when life was shaped exclusively by the plodding forces of evolution. We're standing on the cusp of a new era, one in which we will have primary authority over life's genetic makeup and all its vibrant and varied outputs. Indeed, we are already supplanting the deaf, dumb, and blind system that has shaped genetic material on our planet for eons and replacing it with a conscious, intentional system of human-directed evolution.

That we are unprepared for such colossal responsibility, I have no doubt. But we cannot avoid it. If controlling our own genetic destiny is a terrifying thought, then consider the consequences of having this power but *not* managing to control it. That would be truly terrifying — truly unthinkable.

We must break down the walls that have previously kept science and the public apart and that have encouraged distrust and ignorance to spread unchecked. If anything prevents human beings from rising to the current challenge, it will be these barriers.

My fervent hope is that I can motivate the next generation of scientists to engage much more deeply and openly with the public than my generation has typically done and that they will embrace an ethos of "discussion without dictation" when it comes to deciding how science and technology should be deployed. In this way, scientists can help rebuild the public's trust in us.

There are signs of progress. In recent years, the open-access movement has made many scholarly articles freely available to the public, and the shift toward online courses increases the accessibility of education to students of all ages around the world. These trends are positive, but more needs to be done. Educational institutions need to rethink how students learn and how they can apply their knowledge to societal problems. I'm working to encourage my university, one of the leading public universities in the world, to organize cross-disciplinary meetings, courses, and research projects. By creating opportunities for scientists, writers, psychologists, historians, political scientists, ethicists, economists, and others to work together on real-world problems, we will enhance our collective abilities to explain our work and our disciplines to nonspecialists. I think this will in turn invite students to think more broadly about their fields of expertise and learn how to apply knowledge to problem solving. It's always harder to implement ideas than to formulate them, but I sense a growing interest in such cross-disciplinary initiatives among my colleagues. And in a curious way, CRISPR technology may help spark these efforts due to the many fields that it touches on: science, ethics, economics, sociology, ecology, and evolution.

All scientists, regardless of discipline, need to be prepared to confront the broadest consequences of our work — but we need to communicate its more detailed aspects as well. I was reminded of this at a recent lunch I attended with some of Silicon Valley's greatest technology gurus. One of them said, "Give me ten to twenty million dollars and a team of smart people, and we can solve virtually any engineering challenge." This person obviously knew a thing or two about solving technological problems — a long string of successes attested to that — but ironically, such an approach would not have produced the CRISPR-based gene-editing technology, which was inspired by curiosity-driven research into natural phenomena. The technology we ended up creating did not take anywhere near ten to twenty million dollars to develop, but it *did* require a thorough understanding of the chemistry and biology of bacterial adaptive immunity, a topic that may seem wholly unrelated to gene editing. This is but one example of the importance of fundamental research — the pursuit of science for the sake of understanding our natural world — and its relevance to developing new technologies. Nature, after all, has had a lot more time than humans to conduct experiments!

If there's one overarching point I hope you will take away from this book, it's that humans need to keep exploring the world around us through open-ended scientific research. The wonders of penicillin would never have been discovered had Alexander Fleming not been conducting simple experiments with *Staphylococci* bacteria. Recombinant DNA research — the foundation for modern molecular biology — became possible only with the isolation of DNA-cutting and DNA-copying enzymes from gut- and heat-loving bacteria. Rapid DNA sequencing required experiments on the remarkable properties of bacteria from hot springs. And my colleagues and I would never have created a powerful

gene-editing tool if we hadn't tackled the much more fundamental question of how bacteria fight off viral infections.

The story of CRISPR is a reminder that breakthroughs can come from
unexpected places and that it's important to let a desire to understand
nature dictate the path forward. But it's also a reminder that scientists
and laypeople alike bear a tremendous responsibility for the scientific
process and its outputs. We must continue to support new findings in
all areas of science, and we must wholeheartedly embrace and diligently
exercise our stewardship over these discoveries. For, as history makes
clear, just because we are not ready for scientific progress does not mean
it won't happen. Every time we unlock one of nature's secrets, it signals
the end of one experiment — and the beginning of many others.

— Jennifer Doudna, September 2016

ACKNOWLEDGMENTS

Jennifer & Sam

Writing this book was an exciting and challenging experience for both of us, and it would not have been possible without the generous help and support we received from colleagues, friends, and our families.

We are indebted to our agent, Max Brockman, for promoting our book and enthusiastically supporting the project from the very beginning. We owe a huge thank-you to Alexander Littlefield, our tireless editor at Houghton Mifflin Harcourt, for refining our lengthy (and overly technical) early chapter drafts and for providing creative insights into how best to organize and frame the material; he was a true pleasure to work with. Thanks to everyone else at HMH that played a role in publishing and publicizing our book, especially Pilar Garcia-Brown, Laura Brady, Stephanie Kim, and Michelle Triant. Tracy Roe did a fantastic job copyediting our manuscript (and educating us on the finer points of the *Chicago Manual of Style*), while graciously tolerating all of our last-minute changes. We were extremely lucky to nail down a fantastic artist for the book and are grateful for all the hard work that Jeff Mathison put into bringing to life some tricky scientific concepts with his beautiful ink drawings.

Thanks to Martin Jinek, Blake Wiedenheft, and Jillian Banfield for reading and providing feedback on specific sections of the manuscript. Megan Hochstrasser provided an invaluable service by proofreading the final version of the manuscript. We are also grateful to all the other individuals that contributed to the book through in-person discussions and insightful commentary.

It's an unfortunate fact of science (and indeed, of all academic disciplines) that one cannot possibly acknowledge by name the countless individuals who have contributed to a particular area of research. We have been humbled to be involved in advancing both CRISPR-Cas biology and CRISPR-based gene editing, fields that include many superb scientists. From the early pioneers of gene targeting and gene therapy to the courageous trailblazers of CRISPR biology to the present-day genome engineers, we have been motivated, and continue to be inspired, by the incredible work going on all around us. We hope that interested readers will take the opportunity to share in our excitement by reading some of the many other accounts, articles, and books on CRISPR and gene editing.

Jennifer

I am deeply indebted to my wonderful spouse, Jamie Cate, and to my son, Andrew, for their love, encouragement, and good humor throughout the project and beyond. I could not have completed this work without their support. Although neither of my parents was able to know about the work described in this book, I could never have become a scientist without their belief in me and my passion for discovery. My sisters, Ellen and Sarah, have also been consistent supporters to whom I am indebted. I'm grateful to Rachel Haurwitz, an outstanding scientist who helped me with key parts of the project. Two other scientists, Emmanuelle Charpentier and Jillian Banfield, were central to the early work described in the book, and I thank both of them for the opportunity to work together. I am also extremely appreciative of my assistants, Julie Anderson, Lisa Daitch, and Molly Jorgensen, for 24/7 support and for helping me to juggle other tasks and commitments so that I could carve out time for writing. Finally, I want to thank my coauthor, Sam, for devoting a year of his professional life to this project, which would never have happened without his skillful writing, his scientific insights, and his interest in the broader implications of a transformative technology.

Sam

I owe a big thank-you to Jennifer for having the confidence and trust in me to embark upon this journey together. Ezgi Hacisuleyman was a helpful sounding board and voice of support when I was first considering the project. Blake Wiedenheft, Rebecca Besdin, Annabelle Kleist, Mitchell O'Connell, and Benjamin Oakes provided valuable feedback on the early version of the book proposal. Thanks to Noam Prywes for brainstorming ideas with me over the phone on numerous occasions, and to Sandra Fluck for being so enthusiastic about the project and reading early chapter drafts. Kathryn Quanstrom was a constant source of friendship and support, and she often put up with my complaints when the going got tough. Nofar Hefes sustained me throughout the year of writing and offered me an amazing place to work from. Last but not least, I am indebted to my brother Max and to my parents, Robert Sternberg and Susanne Nimmrichter, without whom I could not have gotten to where I am today. From the first moment I dreamed of writing this book, while vacationing with them in Hawaii, to the last sentence I wrote before submitting the manuscript, they encouraged me through it all, and I can confidently say that this project wouldn't have been possible without their endless love and support.

NOTES

1. THE QUEST FOR A CURE

page

3 *scientists at the National Institutes of Health:* D. H. McDermott et al., "Chromothriptic Cure of WHIM Syndrome," *Cell* 160 (2015): 686–99.

known as WHIM syndrome: WHIM is named after its four major symptomatic manifestations: warts, hypogammaglobulinemia (a deficiency in immunoglobulin), infections, and myelokathexis (a deficiency in certain kinds of white blood cells).

5 *recently discovered phenomenon:* P. J. Stephens et al., "Massive Genomic Rearrangement Acquired in a Single Catastrophic Event During Cancer Development," *Cell* 144 (2011): 27–40.

6 *the scientific literature is peppered with other examples:* R. Hirschhorn, "In Vivo Reversion to Normal of Inherited Mutations in Humans," *Journal of Medical Genetics* 40 (2003): 721–28.

The reason in both cases, scientists determined: R. Hirschhorn et al., "Somatic Mosaicism for a Newly Identified Splice-Site Mutation in a Patient with Adenosine Deaminase-Deficient Immunodeficiency and Spontaneous Clinical Recovery," *American Journal of Human Genetics* 55 (1994): 59–68.

other genetic diseases, such as Wiskott-Aldrich syndrome: B. R. Davis and F. Candotti, "Revertant Somatic Mosaicism in the Wiskott-Aldrich Syndrome," *Immunologic Research* 44 (2009): 127–31.

7 *liver condition called tyrosinemia:* E. A. Kvittingen et al., "Self-Induced Correction of the Genetic Defect in Tyrosinemia Type I," *Journal of Clinical Investigation* 94 (1994): 1657–61.

ichthyosis with confetti: K. A. Choate et al., "Mitotic Recombination in Patients with Ichthyosis Causes Reversion of Dominant Mutations in *KRT10*," *Science* 330 (2010): 94–97.

8 *portmanteau of* gene *and* chromosome: J. Lederberg, "'Ome Sweet 'Omics — A Genealogical Treasury of Words," *Scientist,* April 2, 2001.

17 *"It was clear that we had uncovered":* S. Rogers, "Reflections on Issues Posed by Recombinant DNA Molecule Technology. II," *Annals of the New York Academy of Sciences* 265 (1976): 66–70.

many scientists considered reckless and premature: T. Friedmann and R. Roblin, "Gene Therapy for Human Genetic Disease?," *Science* 175 (1972): 949–55.

the Shope virus didn't even contain an arginase gene: T. Friedmann, "Stanfield Rogers: Insights into Virus Vectors and Failure of an Early Gene Therapy Model," *Molecular Therapy* 4 (2001): 285–88.

23 *that's exactly what had happened:* K. R. Folger et al., "Patterns of Integration of DNA Microinjected into Cultured Mammalian Cells: Evidence for Homologous Recombination Between Injected Plasmid DNA Molecules," *Molecular and Cellular Biology* 2 (1982): 1372–87.

"It will be interesting to determine whether we can exploit": Ibid.

24 *Unbelievably, it worked:* O. Smithies et al., "Insertion of DNA Sequences into the Human Chromosomal Beta-Globin Locus by Homologous Recombination," *Nature* 317 (1985): 230–34.

25 *fix even single mutations:* K. R. Thomas, K. R. Folger, and M. R. Capecchi, "High Frequency Targeting of Genes to Specific Sites in the Mammalian Genome," *Cell* 44 (1986): 419–28.

inactivate them for research purposes: S. L. Mansour, K. R. Thomas, and M. R. Capecchi, "Disruption of the Proto-Oncogene Int-2 in Mouse Embryo-Derived Stem Cells: A General Strategy for Targeting Mutations to Non-Selectable Genes," *Nature* 336 (1988): 348–52.

26 *"Eventually, homologous recombination":* J. Lyon and Peter Gorner, *Altered Fates: Gene Therapy and the Retooling of Human Life* (New York: Norton, 1995), 556.

27 *published a provocative model:* J. W. Szostak et al., "The Double-Strand-Break Repair Model for Recombination," *Cell* 33 (1983): 25–35.

29 *The results of Jasin's experiment:* P. Rouet, F. Smih, and M. Jasin, "Introduction of Double-Strand Breaks into the Genome of Mouse Cells by Expression of a Rare-Cutting Endonuclease," *Molecular and Cellular Biology* 14 (1994): 8096–8106.

32 *Chandrasegaran's chimeric nuclease seemed to work:* Y. G. Kim, J. Cha, and S. Chandrasegaran, "Hybrid Restriction Enzymes: Zinc Finger Fusions to Fok I Cleavage Domain," *Proceedings of the National Academy of Sciences of the United States of America* 93 (1996): 1156–60.

also worked in frog eggs: M. Bibikova et al., "Stimulation of Homologous Recombination Through Targeted Cleavage by Chimeric Nucleases," *Molecular and Cellular Biology* 21 (2001): 289–97.

produce a precise genetic alteration in a whole organism: M. Bibikova et al., "Targeted Chromosomal Cleavage and Mutagenesis in Drosophila Using Zinc-Finger Nucleases," *Genetics* 161 (2002): 1169–75.

Matthew Porteus and David Baltimore were the first: M. H. Porteus and D. Baltimore, "Chimeric Nucleases Stimulate Gene Targeting in Human Cells," *Science* 300 (2003): 763.

Fyodor Urnov and colleagues corrected a mutation: F. D. Urnov et al., "Highly Efficient

Endogenous Human Gene Correction Using Designed Zinc-Finger Nucleases," *Nature* 435 (2005): 646–51.

34 *"But pity the poor TALENs"*: S. Chandrasegaran and D. Carroll, "Origins of Programmable Nucleases for Genome Engineering," *Journal of Molecular Biology* 428 (2016): 963–89.

2. A NEW DEFENSE

42 *But her lab had uncovered an important clue:* G. W. Tyson and J. F. Banfield, "Rapidly Evolving CRISPRs Implicated in Acquired Resistance of Microorganisms to Viruses," *Environmental Microbiology* 10 (2008): 200–207.

43 *pioneering work by Francisco Mojica, a professor in Spain:* F. J. Mojica et al., "Biological Significance of a Family of Regularly Spaced Repeats in the Genomes of Archaea, Bacteria and Mitochondria," *Molecular Microbiology* 36 (2000): 244–46.

Jill pulled three scientific publications, all from 2005: F. J. Mojica et al., "Intervening Sequences of Regularly Spaced Prokaryotic Repeats Derive from Foreign Genetic Elements," *Journal of Molecular Evolution* 60 (2005): 174–82; C. Pourcel, G. Salvignol, and G. Vergnaud, "CRISPR Elements in *Yersinia pestis* Acquire New Repeats by Preferential Uptake of Bacteriophage DNA, and Provide Additional Tools for Evolutionary Studies," *Microbiology* 151 (2005): 653–63; A. Bolotin et al., "Clustered Regularly Interspaced Short Palindrome Repeats (CRISPRs) Have Spacers of Extrachromosomal Origin," *Microbiology* 151 (2005): 2551–61.

Jill's own pioneering research: A. F. Andersson and J. F. Banfield, "Virus Population Dynamics and Acquired Virus Resistance in Natural Microbial Communities," *Science* 320 (2008): 1047–50.

44 *National Institutes of Health team led by Kira Makarova and Eugene Koonin:* K. S. Makarova et al., "A Putative RNA-Interference-Based Immune System in Prokaryotes: Computational Analysis of the Predicted Enzymatic Machinery, Functional Analogies with Eukaryotic RNAi, and Hypothetical Mechanisms of Action," *Biology Direct* 1 (2006): 7.

46 *"dissolved away like sugar in water," vanishing overnight*: D. H. Duckworth, "Who Discovered Bacteriophage?," *Bacteriological Reviews* 40 (1976): 793–802.

the institute had over a thousand employees producing tons of phages a year: C. Zimmer, *A Planet of Viruses* (Chicago: University of Chicago Press, 2011).

20 percent of bacterial infections are treated with phages in Georgia today: G. Naik, "To Fight Growing Threat from Germs, Scientists Try Old-fashioned Killer," *Wall Street Journal,* January 22, 2016.

47 *the first to be synthesized entirely from scratch*: G.P.C. Salmond and P. C. Fineran, "A Century of the Phage: Past, Present and Future," *Nature Reviews Microbiology* 13 (2015): 777–86.

48 *40 percent of all bacteria die every day as a result of deadly phage infections*: F. Rohwer et al., *Life in Our Phage World* (San Diego: Wholon, 2014).

50 *four major bacterial defense systems had been identified:* S. J. Labrie, J. E. Samson, and S. Moineau, "Bacteriophage Resistance Mechanisms," *Nature Reviews Microbiology* 8 (2010): 317–27.

Computational analysis by Ruud Johnson and his colleagues: R. Jansen et al., "Identification of Genes That Are Associated with DNA Repeats in Prokaryotes," *Molecular Microbiology* 43 (2002): 1565–75.

53 *the first bacterium in which a CRISPR sequence had been identified:* Y. Ishino et al., "Nucleotide Sequence of the Iap Gene, Responsible for Alkaline Phosphatase Isozyme Conversion in *Escherichia coli,* and Identification of the Gene Product," *Journal of Bacteriology* 169 (1987): 5429–33.

CRISPR was indeed a bacterial immune system: R. Barrangou et al., "CRISPR Provides Acquired Resistance Against Viruses in Prokaryotes," *Science* 315 (2007): 1709–12.

54 *annual market value of cultures of the bacterium is more than forty billion dollars:* A. Bolotin et al., "Complete Sequence and Comparative Genome Analysis of the Dairy Bacterium *Streptococcus thermophilus,*" *Nature Biotechnology* 22 (2004): 1554–58.

but nothing had solved the problem: M. B. Marcó, S. Moineau, and A. Quiberoni, "Bacteriophages and Dairy Fermentations," *Bacteriophage* 2 (2012): 149–58.

57 *unambiguous evidence that molecules of RNA were involved in CRISPR's antiviral defense:* S.J.J. Brouns et al., "Small CRISPR RNAs Guide Antiviral Defense in Prokaryotes," *Science* 321 (2008): 960–64.

RNA molecules that precisely matched the sequence of CRISPR DNA: T.-H. Tang et al., "Identification of Novel Non-Coding RNAs as Potential Antisense Regulators in the Archaeon *Sulfolobus solfataricus,*" *Molecular Microbiology* 55 (2005): 469–81.

59 *to prove that CRISPR RNAs target the DNA of invading genetic parasites:* L. A. Marraffini and E. J. Sontheimer, "CRISPR Interference Limits Horizontal Gene Transfer in Staphylococci by Targeting DNA," *Science* 322 (2008): 1843–45.

3. CRACKING THE CODE

64 *a protein enzyme called Cas1 had the ability to cut up DNA:* B. Wiedenheft et al., "Structural Basis for DNase Activity of a Conserved Protein Implicated in CRISPR-Mediated Genome Defense," *Structure* 17 (2009): 904–12.

65 *Rachel and Blake found that, like Cas1, it functioned as a chemical cleaver:* R. E. Haurwitz et al., "Sequence- and Structure-Specific RNA Processing by a CRISPR Endonuclease," *Science* 329 (2010): 1355–58.

66 *phage DNA targeted by the CRISPR system got sliced apart:* J. E. Garneau et al., "The CRISPR/Cas Bacterial Immune System Cleaves Bacteriophage and Plasmid DNA," *Nature* 468 (2010): 67–71.

phage eradication in bacteria depended on the presence of specific cas genes: R. Sapranauskas et al., "The *Streptococcus thermophilus* CRISPR/Cas System Provides Immunity in *Escherichia coli,*" *Nucleic Acids Research* 39 (2011): 9275–82.

67 *we obtained the first high-resolution images of the Cascade machine:* B. Wiedenheft et

al., "Structures of the RNA-Guided Surveillance Complex from a Bacterial Immune System," *Nature* 477 (2011): 486–89.

destroyed the viral DNA targeted by Cascade: T. Sinkunas et al., "In Vitro Reconstitution of Cascade-Mediated CRISPR Immunity in *Streptococcus thermophilus*," *EMBO Journal* 32 (2013): 385–94.

68 *nine different types of CRISPR immune systems:* D. H. Haft et al., "A Guild of 45 CRISPR-Associated (Cas) Protein Families and Multiple CRISPR/Cas Subtypes Exist in Prokaryotic Genomes," *PLoS Computational Biology* 1 (2005): e60.

within these basic types there were thought to be ten subtypes: K. S. Makarova et al., "Evolution and Classification of the CRISPR-Cas Systems," *Nature Reviews Microbiology* 9 (2011): 467–77.

two broad classes comprising six types and nineteen subtypes: K. S. Makarova et al., "An Updated Evolutionary Classification of CRISPR-Cas Systems," *Nature Reviews Microbiology* 13 (2015): 722–36; S. Shmakov et al., "Discovery and functional Characterization of Diverse Class 2 CRISPR-Cas Systems," *Molecular Cell* 60 (2015): 385–97.

71 *her paper on the same topic had recently been published:* E. Deltcheva et al., "CRISPR RNA Maturation by Trans-Encoded Small RNA and Host Factor RNase III," *Nature* 471 (2011): 602–7.

73 *it's responsible for over half a million deaths annually:* A. P. Ralph and J. R. Carapetis, "Group A Streptococcal Diseases and Their Global Burden," *Current Topics in Microbiology and Immunology* 368 (2013): 1–27.

85 *"We propose an alternative methodology based on RNA-programmed Cas9":* M. Jinek et al., "A Programmable Dual-RNA-Guided DNA Endonuclease in Adaptive Bacterial Immunity," *Science* 337 (2012): 816–21.

4. COMMAND AND CONTROL

91 *Virginijus Siksnys and colleagues published a similar paper to ours in the fall of 2012:* G. Gasiunas et al., "Cas9-crRNA Ribonucleoprotein Complex Mediates Specific DNA Cleavage for Adaptive Immunity in Bacteria," *Proceedings of the National Academy of Sciences of the United States of America* 109 (2012): 86.

96 *a whopping five articles on CRISPR besides our own:* L. Cong et al., "Multiplex Genome Engineering Using CRISPR/Cas Systems," *Science* 339 (2013): 819–23; P. Mali et al., "RNA-guided Human Genome Engineering via Cas9," *Science* 339 (2013): 823–26; M. Jinek et al., "RNA-programmed Genome Editing in Human Cells," *eLife* 2 (2013): e00471; W. Y. Hwang et al., "Efficient Genome Editing in Zebrafish Using a CRISPR-Cas System," *Nature Biotechnology* 31 (2013): 227–29; S. W. Cho, S. Kim, J. M. Kim and J.-S. Kim, "Targeted Genome Engineering in Human Cells with the Cas9 RNA-guided Endonuclease," *Nature Biotechnology* 31 (2013): 230–32; W. Jiang et al., "RNA-guided Editing of Bacterial Genomes Using CRISPR-Cas Systems," *Nature Biotechnology* 31 (2013): 233–39.

97 *generation of gene-edited mice using CRISPR:* H. Wang et al., "One-Step Generation

of Mice Carrying Mutations in Multiple Genes by CRISPR/Cas-Mediated Genome Engineering," *Cell* 153 (2013): 910–18.

104 *a research team at the University of Texas:* S.-T. Yen et al., "Somatic Mosaicism and Allele Complexity Induced by CRISPR/Cas9 RNA Injections in Mouse Zygotes," *Developmental Biology* 393 (2014): 3–9.

a Japanese research team repeated the same experiment: G. A. Sunagawa et al., "Mammalian Reverse Genetics Without Crossing Reveals *Nr3a* as a Short-Sleeper Gene," *Cell Reports* 14 (2016): 662–77.

109 *Similar research was published by Virginijus Siksnys and colleagues:* Gasiunas et al., "Cas9-crRNA Ribonucleoprotein Complex Mediates Specific DNA Cleavage."

deactivated version of CRISPR had its own uses for manipulating the genome: L. S. Qi et al., "Repurposing CRISPR as an RNA-Guided Platform for Sequence-Specific Control of Gene Expression," *Cell* 152 (2013): 1173–83; L. A. Gilbert et al., "CRISPR-Mediated Modular RNA-Guided Regulation of Transcription in Eukaryotes," *Cell* 154 (2013): 442–51.

111 *this technology would change biotech forever:* M. Herper, "This Protein Could Change Biotech Forever," *Forbes*, March 19, 2013, www.forbes.com/sites/matthew herper/2013/03/19/the-protein-that-could-change-biotech-forever/#7001200f473b.

112 *to researchers in over eighty different countries:* H. Ledford, "CRISPR: Gene Editing Is Just the Beginning," *Nature News*, March 7, 2016.

113 *anyone can set up a CRISPR lab for just $2,000:* K. Loria, "The Process Used to Edit the Genes of Human Embryos Is So Easy You Could Do It in a Community Bio-Hacker Space," *Business Insider*, May 1, 2015.

"everything you need to make precision genome edits in bacteria at home": J. Zayner, "DIY CRISPR Kits, Learn Modern Science by Doing," www.indiegogo.com/projects/diy-crispr-kits-learn-modern-science-by-doing#/.

editing the yeast genome to make new flavors of beer: E. Callaway, "Tapping Genetics for Better Beer," *Nature* 535 (2016): 484–86.

5. THE CRISPR MENAGERIE

119 *discovered gene mutations that made the plant resistant to a pernicious fungus:* P. Piffanelli et al., "A Barley Cultivation-Associated Polymorphism Conveys Resistance to Powdery Mildew," *Nature* 430 (2004): 887–91.

120 *"as plastic in our hands as clay in the hands of the potter":* N. V. Federoff and N. M. Brown, *Mendel in the Kitchen: A Scientist's View of Genetically Modified Foods* (Washington, DC: Joseph Henry Press, 2004), 54.

a German cultivar that had been irradiated with x-rays in 1942: J. H. Jørgensen, "Discovery, Characterization and Exploitation of Mlo Powdery Mildew Resistance in Barley," *Euphytica* 63 (1992): 141–52.

in this case, fungus-resistant barley: R. Büschges et al., "The Barley *Mlo* Gene: A Novel Control Element of Plant Pathogen Resistance," *Cell* 88 (1997): 695–705.

122　*mushrooms that are impervious to browning and premature spoiling:* W. Jiang et al., "Demonstration of CRISPR/Cas9/sgRNA-Mediated Targeted Gene Modification in Arabidopsis, Tobacco, Sorghum and Rice," *Nucleic Acids Research* 41 (2013): e188; N. M. Butler et al., "Generation and Inheritance of Targeted Mutations in Potato (*Solanum Tuberosum* L.) Using the CRISPR/Cas System," *PLoS ONE* 10 (2015): e0144591; S. S. Hall, "Editing the Mushroom," *Scientific American* 314 (2016): 56–63.

to edit the genome of sweet oranges: H. Jia and N. Wang, "Targeted Genome Editing of Sweet Orange Using Cas9/sgRNA," *PLoS ONE* 9 (2014): e93806.

from a bacterial plant disease called huanglongbing: S. Nealon, "Uncoding a Citrus Tree Killer," *UCR Today,* February 9, 2016.

gene editing in bananas can help save the prized Cavendish variety from extinction: D. Cyranoski, "CRISPR Tweak May Help Gene-Edited Crops Bypass Biosafety Regulation," *Nature News,* October 19, 2015.

providing them with a completely new antiviral immune system: A. Chaparro-Garcia, S. Kamoun, and V. Nekrasov, "Boosting Plant Immunity with CRISPR/Cas," *Genome Biology* 16 (2015): 254–57.

an overall fat profile similar to olive oil's: W. Haun et al., "Improved Soybean Oil Quality by Targeted Mutagenesis of the Fatty Acid Desaturase 2 Gene Family," *Plant Biotechnology Journal* 12 (2014): 934–40.

123　*a 70 percent drop in acrylamide levels in potato chips made with the enhanced spuds:* B. M. Clasen et al., "Improving Cold Storage and Processing Traits in Potato Through Targeted Gene Knockout," *Plant Biotechnology Journal* 14 (2016): 169–76.

"the production of heritable improvements in plants or animals": United States Department of Agriculture, "Glossary of Agricultural Biotechnology Terms," last modified February 27, 2013, www.usda.gov/wps/portal/usda/usdahome?navid=BIOTECH_GLOSS&navtype=RT&parentnav=BIOTECH.

124　*94 percent of all soybeans:* USDA Economic Research Service, "Adoption of Genetically Engineered Crops in the U.S.," last modified July 14, 2016, www.ers.usda.gov/data-products/adoption-of-genetically-engineered-crops-in-the-us.aspx.

125　*nearly 60 percent of Americans perceive GMOs as unsafe:* Pew Research Center, "Eating Genetically Modified Foods," www.pewinternet.org/2015/01/29/public-and-scientists-views-on-science-and-society/pi_2015–01–29_science-and-society-03–02/.

126　*delivered to plant cells in this fast-action formulation:* J. W. Woo et al., "DNA-Free Genome Editing in Plants with Preassembled CRISPR-Cas9 Ribonucleoproteins," *Nature Biotechnology* 33 (2015): 1162–64.

the first activist-led protests over the new technology took place in the spring of 2016: "Breeding Controls," *Nature* 532 (2016): 147.

127　*as did some thirty other types of genetically modified plants:* H. Ledford, "Gene-Editing Surges as US Rethinks Regulations," *Nature News,* April 12, 2016.

128　*CRISPR-based plant products will be on the market by the end of the decade:* A. Regalado, "DuPont Predicts CRISPR Plants on Dinner Plates in Five Years," *MIT Technology Review,* October 8, 2015.

revisit the regulation of genetically engineered crops and animals: E. Waltz, "A Face-Lift for Biotech Rules Begins," *Nature Biotechnology* 33 (2015): 1221–22.

the 2016 passage of federal legislation that requires labeling: M. C. Jalonick, "Obama Signs Bill Requiring Labeling of GMO Foods," *Washington Post,* July 29, 2016.

at a cost of over eighty million dollars: C. Harrison, "Going Swimmingly: AquaBounty's GM Salmon Approved for Consumption After 19 Years," *SynBioBeta,* November 23, 2015, http://synbiobeta.com/news/aquabounty-gm-salmon/.

129 *without any changes to its nutritional content or any increased health risks:* A. Pollack, "Genetically Engineered Salmon Approved for Consumption," *New York Times,* November 19, 2015.

a carbon footprint that is around twenty-five times less than for conventional salmon: W. Saletan, "Don't Fear the Frankenfish," *Slate,* November 20, 2015, www.slate.com/articles/health_and_science/science/2015/11/genetically_engineered_aquabounty_salmon_safe_fda_decides.html.

75 percent of respondents wouldn't eat GMO fish: A. Kopicki, "Strong Support for Labeling Modified Foods," *New York Times,* July 27, 2013.

to promise not to sell the salmon: Friends of the Earth, "FDA's Approval of GMO Salmon Denounced," www.foe.org/news/news-releases/2015-11-fdas-approval-of-gmo-salmon-denounced.

the pigs never made it out of the lab: K. Saeki et al., "Functional Expression of a Delta12 Fatty Acid Desaturase Gene from Spinach in Transgenic Pigs," *Proceedings of the National Academy of Sciences of the United States of America* 101 (2004): 6361–66.

to better digest a phosphorus-containing compound called phytate: S. P. Golovan et al., "Pigs Expressing Salivary Phytase Produce Low-Phosphorus Manure," *Nature Biotechnology* 19 (2001): 741–45.

The new breed was finally euthanized in 2012: C. Perkel, "University of Guelph 'Enviropigs' Put Down, Critics Blast 'Callous' Killing," *Huffington Post Canada,* June 21, 2012.

130 *making them a beef producer's dream:* R. Kambadur et al., "Mutations in *Myostatin (GDF8)* in Double-Muscled Belgian Blue and Piedmontese Cattle," *Genome Research* 7 (1997): 910–16.

131 *nature had mirrored previous genetics experiments conducted in mice:* A. C. McPherron, A. M. Lawler, and S. J. Lee, "Regulation of Skeletal Muscle Mass in Mice by a New TGF-β Superfamily Member," *Nature* 387 (1997): 83–90.

Texel sheep, a popular Dutch breed prized for its lean meat: A. Clop et al., "A Mutation Creating a Potential Illegitimate microRNA Target Site in the Myostatin Gene Affects Muscularity in Sheep," *Nature Genetics* 38 (2006): 813–18.

the fastest whippets are in fact the heterozygotes: D. S. Mosher et al., "A Mutation in the Myostatin Gene Increases Muscle Mass and Enhances Racing Performance in Heterozygote Dogs," *PLoS Genetics* 3 (2007): e79.

a team of physicians from Berlin published a remarkable study: M. Schuelke et al., "Myostatin Mutation Associated with Gross Muscle Hypertrophy in a Child," *New England Journal of Medicine* 350 (2004): 2682–88.

132 *editing the* myostatin *gene in normal individuals to unleash enhanced, superhuman strength:* E. P. Zehr, "The Man of Steel, Myostatin, and Super Strength," *Scientific American,* June 14, 2013.

133 *gene-edited pigs had over 10 percent more lean meat than their unedited counterparts:* L. Qian et al., "Targeted Mutations in *Myostatin* by Zinc-Finger Nucleases Result in Double-Muscled Phenotype in Meishan Pigs," *Scientific Reports* 5 (2015): 14435.

The scientists performed gene editing in a breed of goats known as Shannbei: X. Wang et al., "Generation of Gene-Modified Goats Targeting *MSTN* and *FGF5* via Zygote Injection of CRISPR/Cas9 System," *Scientific Reports* 5 (2015): 13878.

134 *so that chickens produce only females:* S. Reardon, "Welcome to the CRISPR Zoo," *Nature News,* March 9, 2016.

porcine genomes are being modified so that pigs can be fattened with less food: A. Harmon, "Open Season Is Seen in Gene Editing of Animals," *New York Times,* November 26, 2015.

similar strategies have been proposed to remove allergens in cow milk: C. Whitelaw et al., "Genetically Engineering Milk," *Journal of Dairy Research* 83 (2016): 3–11.

heavy price is paid by the animals themselves: D. J. Holtkamp et al., "Assessment of the Economic Impact of Porcine Reproductive and Respiratory Syndrome Virus on United States Pork Producers," *Journal of Swine Health and Production* 21 (2013): 72–84.

After using CRISPR to create gene-knockout pigs: K. M. Whitworth et al., "Use of the CRISPR/Cas9 System to Produce Genetically Engineered Pigs from In Vitro–Derived Oocytes and Embryos," *Biology of Reproduction* 91 (2014): 1–13.

135 *the gene-edited pigs remained completely healthy and free of any traces of virus:* K. M. Whitworth et al., "Gene-Edited Pigs Are Protected from Porcine Reproductive and Respiratory Syndrome Virus," *Nature Biotechnology* 34 (2016): 20–22.

even more deadly, with some strains causing near 100 percent mortality: Center for Food Security and Public Health, "African Swine Fever," www.cfsph.iastate.edu/Fact sheets/pdfs/african_swine_fever.pdf.

the UK team zeroed in on a single gene that seemed to explain their remarkable resistance: C. J. Palgrave et al., "Species-Specific Variation in RELA Underlies Differences in NF-κB Activity: A Potential Role in African Swine Fever Pathogenesis," *Journal of Virology* 85 (2011): 6008–14.

so the scientists simply edited the domestic pigs' genes to match it: S. G. Lillico et al., "Mammalian Interspecies Substitution of Immune Modulatory Alleles by Genome Editing," *Scientific Reports* 6 (2016): 21645.

a tiny refinement that ultimately produces healthier animals: H. Devlin, "Could These Piglets Become Britain's First Commercially Viable GM Animals?," *Guardian,* June 23, 2015.

136 *a significant amount of stress and pain for the traumatized calves:* B. Graf and M. Senn, "Behavioural and Physiological Responses of Calves to Dehorning by Heat Cauteriza-

tion with or Without Local Anaesthesia," *Applied Animal Behaviour Science* 62 (1999): 153–71.

137 *a German research team discovered the exact genetic cause:* I. Medugorac et al., "Bovine Polledness — an Autosomal Dominant Trait with Allelic Heterogeneity," *PLoS ONE* 7 (2012): e39477.

the scientists at Recombinetics used gene editing to copy the exact same change: D. F. Carlson et al., "Production of Hornless Dairy Cattle from Genome-Edited Cell Lines," *Nature Biotechnology* 34 (2016): 479–81.

two hornless dairy calves named Spotigy and Buri: K. Grens, "GM Calves Move to University," *Scientist,* December 21, 2015.

138 *there are well over thirty thousand unique mouse strains in existence:* N. Rosenthal and Steve Brown, "The Mouse Ascending: Perspectives for Human-Disease Models," *Nature Cell Biology* 9 (2007): 993–99; www.findmice.org/repository.

a Chinese team created gene-edited cynomolgus monkeys: B. Shen et al., "Generation of Gene-Modified Cynomolgus Monkey via Cas9/RNA-Mediated Gene Targeting in One-Cell Embryos," *Cell* 156 (2014): 836–43.

139 *a gene that is mutated in over 50 percent of human cancers:* H. Wan et al., "One-Step Generation of *p53* Gene Biallelic Mutant Cynomolgus Monkey via the CRISPR/Cas System," *Cell Research* 25 (2015): 258–61.

mutations that cause Duchenne muscular dystrophy: Y. Chen et al., "Functional Disruption of the Dystrophin Gene in Rhesus Monkey Using CRISPR/Cas9," *Human Molecular Genetics* 24 (2015): 3764–74.

being exploited to target genes implicated in neural disorders: Z. Tu et al., "CRISPR/Cas9: A Powerful Genetic Engineering Tool for Establishing Large Animal Models of Neurodegenerative Diseases," *Molecular Neurodegeneration* 10 (2015): 35–42; Z. Liu et al., "Generation of a Monkey with *MECP2* Mutations by TALEN-Based Gene Targeting," *Neuroscience Bulletin* 30 (2014): 381–86.

140 *drug that is purified from the egg whites of transgenic chickens:* C. Sheridan, "FDA Approves 'Farmaceutical' Drug from Transgenic Chickens," *Nature Biotechnology* 34 (2016): 117–19.

There are numerous benefits to extracting the drugs from transgenic animals: L. R. Bertolini et al., "The Transgenic Animal Platform for Biopharmaceutical Production," *Transgenic Research* 25 (2016): 329–43.

CRISPR enables outright replacement of pig genes with their human gene counterparts: J. Peng et al., "Production of Human Albumin in Pigs Through CRISPR/Cas9-Mediated Knockin of Human cDNA into Swine Albumin Locus in the Zygotes," *Scientific Reports* 5 (2015): 16705.

more than 124,000 patients are currently on the waiting list for transplants: D. Cooper et al., "The Role of Genetically Engineered Pigs in Xenotransplantation Research," *Journal of Pathology* 238 (2016): 288–99.

a new individual is added to the national transplant list every ten minutes: U.S. Department of Health and Human Services, "The Need Is Real: Data," www.organdonor.gov/about/data.html.

141 *to eliminate the risk that porcine viruses embedded in the pig genome:* L. Yang et al., "Genome-Wide Inactivation of Porcine Endogenous Retroviruses (PERVs)," *Science* 350 (2015): 1101–4.

the goal is to provide "an unlimited supply of transplantable organs": A. Regalado, "Surgeons Smash Records with Pig-to-Primate Organ Transplants," *MIT Technology Review,* August 12, 2015.

142 *a brand-new breed of miniaturized pig — the so-called micropig:* D. Cyranoski, "Gene-Edited 'Micropigs' to Be Sold as Pets at Chinese Institute," *Nature News,* September 29, 2015.

CRISPR in micropigs to generate a human Parkinson's disease model: X. Wang et al., "One-Step Generation of Triple Gene-Targeted Pigs Using CRISPR/Cas9 System," *Scientific Reports* 6 (2016): 20620.

143 *"for the sole purpose of satisfying idiosyncratic aesthetic preferences of humans":* Cyranoski, "Gene-Edited 'Micropigs.'"

Cavalier King Charles spaniels suffer from seizures and persistent pain due to their deformed skulls: C. Maldarelli, "Although Purebred Dogs Can Be Best in Show, Are They Worst in Health?," *Scientific American,* February 21, 2014.

The two puppies that contained the intended mutations were named Hercules and Tiangou: Q. Zou et al., "Generation of Gene-Target Dogs Using CRISPR/Cas9 System," *Journal of Molecular Cell Biology* 7 (2015): 580–83.

he noted the potential advantages of extra muscle for police and military applications: A. Regalado, "First Gene-Edited Dogs Reported in China," *MIT Technology Review,* October 19, 2015.

144 *CRISPR to generate a bizarre array of bodily transformations in crustaceans:* A. Martin et al., "CRISPR/Cas9 Mutagenesis Reveals Versatile Roles of Hox Genes in Crustacean Limb Specification and Evolution," *Current Biology* 26 (2016): 14–26.

CRISPR might be used to create mythical creatures like winged dragons: M. Evans, "Could Scientists Create Dragons Using CRISPR Gene Editing?," *BBC News,* January 3, 2016.

"a very large reptile that looks at least somewhat like the European or Asian dragon": R. A. Charo and H. T. Greely, "CRISPR Critters and CRISPR Cracks," *American Journal of Bioethics* 15 (2015): 11–17.

This strategy is being undertaken in Europe to bring back the aurochs: B. Switek, "How to Resurrect Lost Species," *National Geographic News,* March 11, 2013; S. Blakeslee, "Scientists Hope to Bring a Galápagos Tortoise Species Back to Life," *New York Times,* December 14, 2015.

145 *the scientists achieved the first-ever resurrection of an extinct animal:* J. Folch et al., "First Birth of an Animal from an Extinct Subspecies *(Capra pyrenaica pyrenaica)* by Cloning," *Theriogenology* 71 (2009): 1026–34.

to resurrect woolly mammoths: K. Loria and D. Baer, "Korea's Radical Cloning Lab Told

Us About Its Breathtaking Plan to Bring Back the Mammoth," *Tech Insider,* September 10, 2015.

the 1,668 genes that differ between the two genomes: V. J. Lynch et al., "Elephantid Genomes Reveal the Molecular Bases of Woolly Mammoth Adaptations to the Arctic," *Cell Reports* 12 (2015): 217–28.

Church's team used CRISPR to convert the elephant variant to the woolly mammoth variant: J. Leake, "Science Close to Creating a Mammoth," *Sunday Times,* March 22, 2015.

146 *would it simply be an elephant with new traits inspired by woolly mammoth genetics:* B. Shapiro, "Mammoth 2.0: Will Genome Engineering Resurrect Extinct Species?," *Genome Biology* 16 (2015): 228–30.

"enhance biodiversity through the genetic rescue of endangered and extinct species": Long Now Foundation, "What We Do," http://reviverestore.org/what-we-do/.

148 *evolutionary biologist Austin Burt proposed a way to harness selfish genes:* A. Burt, "Site-Specific Selfish Genes as Tools for the Control and Genetic Engineering of Natural Populations," *Proceedings of the Royal Society of London B* 270 (2003): 921–28.

George Church's team at Harvard, led by Kevin Esvelt, proposed a way: K. M. Esvelt et al., "Concerning RNA-Guided Gene Drives for the Alteration of Wild Populations," *eLife* 3 (2014): e03401.

149 *reported the first successful demonstration of a CRISPR gene drive:* V. M. Gantz and E. Bier, "The Mutagenic Chain Reaction: A Method for Converting Heterozygous to Homozygous Mutations," *Science* 348 (2015): 442–44.

150 *spread a gene that gave the offspring resistance to* Plasmodium falciparum: V. M. Gantz et al., "Highly Efficient Cas9-Mediated Gene Drive for Population Modification of the Malaria Vector Mosquito *Anopheles stephensi,*" *Proceedings of the National Academy of Sciences of the United States of America* 112 (2015): E6736–43.

highly transmissive CRISPR gene drives that spread genes for female sterility: A. Hammond et al., "A CRISPR-Cas9 Gene Drive System Targeting Female Reproduction in the Malaria Mosquito Vector *Anopheles gambiae,*" *Nature Biotechnology* 34 (2016): 78–83.

eliminated certain agricultural pests through North and Central America: L. Alphey et al., "Sterile-Insect Methods for Control of Mosquito-Borne Diseases: An Analysis," *Vector Borne and Zoonotic Diseases* 10 (2010): 295–311.

field trials have already commenced in Malaysia, Brazil, and Panama: L. Alvarez, "A Mosquito Solution (More Mosquitoes) Raises Heat in Florida Keys," *New York Times,* February 19, 2015.

151 *it would have spread genes encoding CRISPR, along with the yellow-body trait:* "Gene Intelligence," *Nature* 531 (2016): 140.

the importance of developing guidelines that ensure future research proceeds safely: O. S. Akbari et al., "Biosafety: Safeguarding Gene Drive Experiments in the Laboratory," *Science* 349 (2015): 927–29.

One of these is the so-called reversal drive: J. E. DiCarlo et al., "Safeguarding CRISPR-Cas9 Gene Drives in Yeast," *Nature Biotechnology* 33 (2015): 1250–55.

a recent report authored by the National Academies of Sciences: National Academies of Sciences, Engineering, and Medicine, "Gene Drives on the Horizon: Advancing Science, Navigating Uncertainty, and Aligning Research with Public Values," http://nas-sites.org/gene-drives/.

152 *could be militarized and weaponized:* ETC Group, "Stop the Gene Bomb! ETC Group Comment on NAS Report on Gene Drives," June 8, 2016, www.etcgroup.org/content/stop-gene-bomb-etc-group-comment-nas-report-gene-drives.

"Clearly, the technology described here is not to be used lightly": A. Burt, "Site-Specific Selfish Genes as Tools for the Control and Genetic Engineering of Natural Populations," *Proceedings of the Royal Society of London B* 270 (2003): 921–28.

stamping out infectious diseases such as Lyme disease, which is caused by certain bacteria transmitted by ticks: B. J. King, "Are Genetically Engineered Mice the Answer to Combating Lyme Disease?," NPR, June 16, 2016.

have an annual death toll in excess of one million: American Mosquito Control Association, "Mosquito-Borne Diseases," www.mosquito.org/mosquito-borne-diseases.

153 *"If we eradicated them tomorrow":* J. Fang, "Ecology: A World Without Mosquitoes," *Nature* 466 (2010): 432–34.

6. TO HEAL THE SICK

155 *Three startup therapeutics companies:* The three companies are Editas Medicine, Intellia Therapeutics, and CRISPR Therapeutics.

156 *The first clinical trial using CRISPR to be approved:* S. Reardon, "First CRISPR Clinical Trial Gets Green Light from US Panel," *Nature News,* June 22, 2016.

A new biotech institute in San Francisco: Y. Anwar, "UC Berkeley to Partner in $600M Chan Zuckerberg Science 'Biohub,'" *Berkeley News,* September 21, 2016.

launching the Innovative Genomics Institute: R. Sanders, "New DNA-Editing Technology Spawns Bold UC Initiative," *Berkeley News,* March 18, 2014.

first outright, CRISPR-based cure: Y. Wu et al., "Correction of a Genetic Disease in Mouse via Use of CRISPR-Cas9," *Cell Stem Cell* 13 (2013): 659–62.

164 *1 to 2 percent of Caucasians worldwide (most of them in northeastern Europe) are fortunate enough to have this trait:* K. Allers and T. Schneider, "CCR5Δ32 Mutation and HIV Infection: Basis for Curative HIV Therapy," *Current Opinion in Virology* 14 (2015): 24–29.

165 *reduced risk of certain inflammatory diseases:* S. G. Deeks and J. M. McCune, "Can HIV Be Cured with Stem Cell Therapy?," *Nature Biotechnology* 28 (2010): 807–10.

increase in susceptibility to the West Nile virus: W. G. Glass et al., "CCR5 Deficiency Increases Risk of Symptomatic West Nile Virus Infection," *Journal of Experimental Medicine* 203 (2006): 35–40.

Sangamo researchers conducted a clinical trial: P. Tebas et al., "Gene Editing of *CCR5* in Autologous CD4 T Cells of Persons Infected with HIV," *New England Journal of Medicine* 370 (2014): 901–10.

166 *"are safe within the limits of this study"*: Ibid.

169 *the ravages of the disease could be reversed*: N. Wade, "Gene Editing Offers Hope for Treating Duchenne Muscular Dystrophy, Studies Find," *New York Times,* December 31, 2015.

170 *to cure mice of a genetic mutation that causes a condition known as tyrosinemia*: H. Yin et al., "Therapeutic Genome Editing by Combined Viral and Non-Viral Delivery of CRISPR System Components in Vivo," *Nature Biotechnology* 34 (2016): 328–33.
each with its unique set of advantages and disadvantages: X. Chen and M.A.F.V. Gonçalves, "Engineered Viruses as Genome Editing Devices," *Molecular Therapy* 24 (2015): 447–57.

172 *half a million people die from cancer every year*: American Cancer Society, *Cancer Facts and Figures 2016* (Atlanta: American Cancer Society, 2016).

173 *to understand the genetic causes of acute myeloid leukemia*: D. Heckl et al., "Generation of Mouse Models of Myeloid Malignancy with Combinatorial Genetic Lesions Using CRISPR-Cas9 Genome Editing," *Nature Biotechnology* 32 (2014): 941–46.

174 *one of the first to pioneer such a genome-wide knockout screen*: T. Wang et al., "Identification and Characterization of Essential Genes in the Human Genome," *Science* 350 (2015): 1096–1101.

176 *the first human whose life was saved by therapeutic gene editing*: S. Begley, "Medical First: Gene-Editing Tool Used to Treat Girl's Cancer," *STAT News,* November 5, 2015; A. Pollack, "A Cell Therapy Untested in Humans Saves a Baby with Cancer," *New York Times,* November 5, 2015.
Layla's condition had not improved despite chemotherapy: W. Qasim et al., "First Clinical Application of TALEN Engineered Universal CAR19 T Cells in B-ALL," paper presented at the annual meeting for the American Society of Hematology, Orlando, Florida, December 5–8, 2015.

177 *inject human patients with cells that had been modified using CRISPR*: D. Cyranoski, "CRISPR Gene-Editing Tested in a Person for the First Time," *Nature News,* November 15, 2016.

178 *the Cas9 enzyme would in some cases still cut the DNA*: M. Jinek et al., "A Programmable Dual-RNA-Guided DNA Endonuclease in Adaptive Bacterial Immunity," *Science* 337 (2012): 816–21.
together with a team from Harvard University led by David Liu: V. Pattanayak et al., "High-Throughput Profiling of Off-Target DNA Cleavage Reveals RNA-Programmed Cas9 Nuclease Specificity," *Nature Biotechnology* 31 (2013): 839–43.

179 *Other labs also conducted similar experiments inside cells*: Y. Fu et al., "High-Frequency Off-Target Mutagenesis Induced by CRISPR-Cas Nucleases in Human Cells," *Nature Biotechnology* 31 (2013): 822–26; P. D. Hsu et al., "DNA Targeting Specificity of RNA-Guided Cas9 Nucleases," *Nature Biotechnology* 31 (2013): 827–32.

180 *have developed higher-fidelity versions of CRISPR*: F. Urnov, "Genome Editing: The Domestication of Cas9," *Nature* 529 (2016): 468–69.

7. THE RECKONING

187 *bringing the steady march of CRISPR research right to* Homo sapiens' *evolutionary front door:* B. Shen et al., "Generation of Gene-Modified Cynomolgus Monkey via Cas9/RNA-Mediated Gene Targeting in One-Cell Embryos," *Cell* 156.

189 *"power to shape his own biologic destiny":* M. W. Nirenberg, "Will Society Be Prepared?," *Science* 157 (1967): 633.
"potentially one of the most important concepts to arise in the history of mankind": R. L. Sinsheimer, "The Prospect for Designed Genetic Change," *American Scientist* 57 (1969): 134–42.

190 *"might be like the young boy who loves to take things apart":* W. F. Anderson, "Genetics and Human Malleability," *Hastings Center Report* 20 (1990): 21–24.

193 *records of the conference helped reassure me years later:* G. Stock and J. Campbell, eds., *Engineering the Human Germline: An Exploration of the Science and Ethics of Altering the Genes We Pass to Our Children* (Oxford: Oxford University Press, 2000).
A report authored a few years later by the American Association for the Advancement of Science: M. S. Frankel and A. R. Chapman, *Human Inheritable Genetic Modifications: Assessing Scientific, Ethical, Religious, and Policy Issues* (Washington, DC: American Association for the Advancement of Science, 2000).
the Genetics and Public Policy Center reached similar conclusions: S. Baruch, *Human Germline Genetic Modification: Issues and Options for Policymakers* (Washington, DC: Genetics and Public Policy Center, 2005).

196 *the first country in the world to approve regulations permitting its clinical use:* J. Schandera and T. K. Mackey, "Mitochondrial Replacement Techniques: Divergence in Global Policy," *Trends in Genetics* 32 (2016): 385–90.
recommended that the Food and Drug Administration approve future trials of three-parent IVF: S. Reardon, "US Panel Greenlights Creation of Male 'Three-Person' Embryos," *Nature News,* February 3, 2016.

199 *"I want to understand the uses and implications of this amazing technology you've developed":* I first described this dream in an interview with Michael Specter, who wrote about it in a November 2015 feature story on CRISPR in the *New Yorker.*
had already been shipped to dozens of countries: J. K. Joung, D. F. Voytas, and J. Kamens, "Accelerating Research Through Reagent Repositories: The Genome Editing Example," *Genome Biology* 16 (2015): 255–58.
protocols needed to create designer mutations in mammals: Shen, "Generation of Gene-Modified Cynomolgus Monkey."
also sold online to any consumer with a hundred dollars: J. Zayner, "DIY CRISPR Kits, Learn Modern Science by Doing," www.indiegogo.com/projects/diy-crispr-kits-learn-modern-science-by-doing#/.
biohackers messing with more complex genetic systems: P. Skerrett, "Is Do-It-Yourself CRISPR as Scary as It Sounds?," *STAT News,* March 14, 2016.

200 *"It is my judgment in these things that when you see something that is technically sweet":*

United States Atomic Energy Commission, *In the Matter of J. Robert Oppenheimer: Transcript of Hearing Before Personnel Security Board*, vol. 2 (Washington, DC: GPO, 1954), www.osti.gov/includes/opennet/includes/Oppenheimer%20hearings/Vol%20 II%20Oppenheimer.pdf.

202 *he did so by combining DNA from three sources:* D. A. Jackson, R. H. Symons, and P. Berg, "Biochemical Method for Inserting New Genetic Information into DNA of Simian Virus 40: Circular SV40 DNA Molecules Containing Lambda Phage Genes and the Galactose Operon of *Escherichia coli*," *Proceedings of the National Academy of Sciences of the United States of America* 69 (1972): 2904–9.

203 *a notable report titled "Potential Biohazards of Recombinant DNA Molecules":* P. Berg et al., "Letter: Potential Biohazards of Recombinant DNA Molecules," *Science* 185 (1974): 303.

Much has been written about Asilomar II: Institute of Medicine (US) Committee to Study Decision Making; K. E. Hanna, ed., *Biomedical Politics* (Washington, DC: National Academies Press, 1991); M. Rogers, *Biohazard* (New York: Knopf, 1977); P. Berg and M. F. Singer, "The Recombinant DNA Controversy: Twenty Years Later," *Proceedings of the National Academy of Sciences of the United States of America* 92 (1995): 9011–13.

204 *Berg and his colleagues decided that most experiments should proceed:* P. Berg et al., "Asilomar Conference on Recombinant DNA Molecules," *Science* 188 (1975): 991–94.

gave rise to a consensus that allowed research to proceed with popular support: P. Berg, "Meetings That Changed the World: Asilomar 1975: DNA Modification Secured," *Nature* 455 (2008): 290–91.

the meeting failed to cast a wide enough net outside the scientific community: "After Asilomar," *Nature* 526 (2015): 293–94.

topics like biosecurity and ethics from the meeting's agenda: S. Jasanoff, J. B. Hurlbut, and K. Saha, "CRISPR Democracy: Gene Editing and the Need for Inclusive Deliberation," *Issues in Science and Technology* 32 (2015).

"This approach gets democracy wrong": J. B. Hurlbut, "Limits of Responsibility: Genome Editing, Asilomar, and the Politics of Deliberation," *Hastings Center Report* 45 (2015): 11–14.

205 *creation of a governmental authority known as the Recombinant DNA Advisory Committee:* N. A. Wivel, "Historical Perspectives Pertaining to the NIH Recombinant DNA Advisory Committee," *Human Gene Therapy* 25 (2014): 19–24.

211 *"A Prudent Path Forward for Genomic Engineering and Germline Gene Modification":* D. Baltimore et al., "Biotechnology: A Prudent Path Forward for Genomic Engineering and Germline Gene Modification," *Science* 348 (2015): 36–38.

212 *The New York Times ran a front-page story that generated hundreds of reader comments:* N. Wade, "Scientists Seek Ban on Method of Editing the Human Genome," *New York Times*, March 19, 2015.

our perspective was also picked up by media outlets: R. Stein, "Scientists Urge Temporary Moratorium on Human Genome Edits," *All Things Considered*, NPR, March 20,

2015; "Scientists Right to Pause for Genetic Editing Discussion," *Boston Globe,* March 23, 2015.

team writing in the journal Nature *had called for a ban on germline editing:* E. Lanphier et al., "Don't Edit the Human Germline," *Nature* 519 (2015): 410–11.

MIT Technology Review *had recently published a riveting piece on germline editing:* A. Regalado, "Engineering the Perfect Baby," *MIT Technology Review,* March 5, 2015.

8. WHAT LIES AHEAD

214 *The article, published in the journal* Protein and Cell: P. Liang et al., "CRISPR/ Cas9-Mediated Gene Editing in Human Tripronuclear Zygotes," *Protein and Cell* 6 (2015): 363–72.

215 *"pressing need to further improve the fidelity and specificity of the CRISPR/Cas9 platform":* Ibid.

216 *had fully complied with existing regulations in China:* X. Zhai, V. Ng, and R. Lie, "No Ethical Divide Between China and the West in Human Embryo Research," *Developing World Bioethics* 16 (2016): 116–20.

partly because they had ethical objections to the experiments it described: D. Cyranoski and S. Reardon, "Chinese Scientists Genetically Modify Human Embryos," *Nature News,* April 22, 2015.

"the sort of deranged motivation that sometimes prompts people to do things": G. Kolata, "Chinese Scientists Edit Genes of Human Embryos, Raising Concerns," *New York Times,* April 23, 2015.

"strong stance against gene editing in, or gene modification of, human cells": T. Friedmann et al., "ASGCT and JSGT Joint Position Statement on Human Genomic Editing," *Molecular Therapy* 23 (2015): 1282.

217 *"a moratorium on any clinical application of gene editing human embryos is critical":* R. Jaenisch, "A Moratorium on Human Gene Editing to Treat Disease Is Critical," *Time,* April 23, 2015.

"the Administration believes that altering the human germline for clinical purposes": J. Holdren, "A Note on Genome Editing," May 26, 2015, www.whitehouse.gov/ blog/2015/05/26/note-genome-editing.

the NIH would not provide governmental funding for any research: Francis S. Collins, "Statement on NIH Funding of Research Using Gene-Editing Technologies in Human Embryos," April 29, 2015, www.nih.gov/about-nih/who-we-are/nih-director/state ments/statement-nih-funding-research-using-gene-editing-technologies-human -embryos.

genome editing as one of the six weapons of mass destruction and proliferation: J. R. Clapper, "Worldwide Threat Assessment of the US Intelligence Community," February 9, 2016, www.dni.gov/files/documents/SASC_Unclassified_2016_ATA_ SFR_FINAL.pdf.

218 *"Research into gene-editing is not an option, it is a moral necessity":* J. Savulescu et al.,

"The Moral Imperative to Continue Gene Editing Research on Human Embryos," *Protein and Cell* 6 (2015): 476–79.

"the primary moral goal for today's bioethics can be summarized in a single sentence": S. Pinker, "The Moral Imperative for Bioethics," *Boston Globe,* August 1, 2015.

219 *their statement on genetic modification of the human germline, the Hinxton Group:* Hinxton Group, "Statement on Genome Editing Technologies and Human Germline Genetic Modification," September 3, 2015, www.hinxtongroup.org/Hinxton2015_Statement.pdf.

rumors that multiple other Chinese groups were already planning or performing: Cyranoski and Reardon, "Chinese Scientists."

scientists at the prestigious Francis Crick Institute in London: D. Cressey, A. Abbott, and H. Ledford, "UK Scientists Apply for License to Edit Genes in Human Embryos," *Nature News,* September 18, 2015.

221 *experts from a wide range of fields:* For a complete list, see the National Academies of Sciences, Engineering, and Medicine, "International Summit on Human Gene Editing," December 1–3, 2015, www.nationalacademies.org/gene-editing/Gene-Edit-Summit/index.htm.

223 *somewhere between two and ten novel DNA mutations creep into the genome:* I. Martincorena and P. J. Campbell, "Somatic Mutation in Cancer and Normal Cells," *Science* 34 (2015): 1483–89.

Every person experiences roughly one million mutations throughout the body: M. Porteus, "Therapeutic Genome Editing of Hematopoietic Cells," Presentation at Inserm Workshop 239, CRISPR-Cas9: Breakthroughs and Challenges, Bordeaux, France, April 6–8, 2016.

every single letter of the genome will have been mutated at least once: M. Lynch, "Rate, Molecular Spectrum, and Consequences of Human Mutation," *Proceedings of the National Academy of Sciences of the United States of America* 107 (2010): 961–68.

"Genetic editing would be a droplet in the maelstrom of naturally churning genomes": S. Pinker in P. Skerrett, "Experts Debate: Are We Playing with Fire When We Edit Human Genes?," *STAT News,* November 17, 2015.

224 *research in mice has demonstrated that eggs and sperm can be grown in the laboratory:* Q. Zhou et al., "Complete Meiosis from Embryonic Stem Cell-Derived Germ Cells In Vitro," *Cell Stem Cell* 18 (2016): 330–40; K. Morohaku et al., "Complete In Vitro Generation of Fertile Oocytes from Mouse Primordial Germ Cells," *Proceedings of the National Academy of Sciences of the United States of America* 113 (2016): 9021–26.

make the resulting human immune to HIV but more susceptible to the West Nile virus: J. K. Lim et al., "CCR5 Deficiency Is a Risk Factor for Early Clinical Manifestations of West Nile Virus Infection but Not for Viral Transmission," *Journal of Infectious Diseases* 201 (2010): 178–85.

rid them of the disease but also deprive them of the mutation's protection against malaria: M. Aidoo et al., "Protective Effects of the Sickle Cell Gene Against Malaria Morbidity and Mortality," *Lancet* 359 (2002): 1311–12.

225 *people who carry one copy of the mutated gene that causes cystic fibrosis:* E. M. Poolman and A. P. Galvani, "Evaluating Candidate Agents of Selective Pressure for Cystic Fibrosis," *Journal of the Royal Society* 4 (2007): 91–98.

Even gene variants implicated in neurodegenerative diseases like Alzheimer's may have benefits: E. S. Lander, "Brave New Genome," *New England Journal of Medicine* 373 (2015): 5–8.

"The notion that we need complete knowledge of the whole human genome": G. Church, "Should Heritable Gene Editing Be Used on Humans?," *Wall Street Journal,* April 10, 2016.

227 *2016 Pew Research poll found that 50 percent of adults in the U.S. oppose the idea:* C. Funk, B. Kennedy, and E. P. Sciupac, *U.S. Public Opinion on the Future Use of Gene Editing* (Washington, DC: Pew Research Center, 2016); "Genetic Modifications for Babies," Pew Research Center, January 28, 2015, www.pewinternet.org/2015/01/29/public -and-scientists-views-on-science-and-society/pi_2015–01–29_science-and-society -03–25.

others welcome human involvement in the works of nature as long as the goals: D. Carroll and R. A. Charo, "The Societal Opportunities and Challenges of Genome Editing," *Genome Biology* 16 (2015): 242–50.

"Evolution has been working toward optimizing the human genome for 3.85 billion years": Skerrett, "Experts Debate."

229 *"the human genome underlies the fundamental unity of all members of the human family":* United Nations Educational, Scientific and Cultural Organization, "Universal Declaration on the Human Genome and Human Rights," November 11, 1997, www.unesco.org/new/en/social-and-human-sciences/themes/bioethics/human-ge nome-and-human-rights/.

"jeopardize the inherent and therefore equal dignity of all human beings and renew eugenics": United Nations Educational, Scientific and Cultural Organization, "Report of the IBC on Updating Its Reflection on the Human Genome and Human Rights," October 2, 2015, http://unesdoc.unesco.org/images/0023/002332/233258E.pdf.

Some bioethicists have voiced similar concerns: G. Annas, "Viewpoint: Scientists Should Not Edit Genomes of Human Embryos," April 30, 2015, www.bu.edu/sph/2015/04/30/ scientists-should-not-edit-genomes-of-human-embryos/.

231 *Recent gene therapies that have hit the market come with a price tag:* E. C. Hayden, "Promising Gene Therapies Pose Million-Dollar Conundrum," *Nature News,* June 15, 2016; S. H. Orkin and P. Reilly, "Medicine: Paying for Future Success in Gene Therapy," *Science* 352 (2016): 1059–61.

233 *As disability-rights advocates have pointed out:* T. Shakespeare, "Gene Editing: Heed Disability Views," *Nature* 527 (2015): 446.

234 *That's because* eugenic, *as it was originally defined, means "well-born":* C. J. Epstein, "Is Modern Genetics the New Eugenics?," *Genetics in Medicine* 5 (2003): 469–75.

"Anyone who has to actually face the reality of one of these diseases": E. C. Hayden, "Should You Edit Your Children's Genes?," *Nature News,* February 23, 2016.

235 *current government regulations on the topic are variable:* M. Araki and T. Ishii, "International Regulatory Landscape and Integration of Corrective Genome Editing into In Vitro Fertilization," *Reproductive Biology and Endocrinology* 12 (2014): 108–19.

236 *"which result in modifications to the subject's germline genetic identity":* R. Isasi, E. Kleiderman, and B. M. Knoppers, "Editing Policy to Fit the Genome?," *Science* 351 (2016): 337–39.

237 *"in which a human embryo is intentionally created or modified":* I. G. Cohen and E. Y. Adashi, "The FDA Is Prohibited from Going Germline," *Science* 353 (2016): 545–46.
Medical tourists have already spent millions of dollars: D.B.H. Mathews et al., "CRISPR: A Path Through the Thicket," *Nature* 527 (2015): 159–61.
gene therapy treatments to increase muscle mass and lengthen lifespan: A. Regalado, "A Tale of Do-It-Yourself Gene Therapy," *MIT Technology Review,* October 14, 2015.

238 *Some authors have predicted:* G. O. Schaefer, "The Future of Genetic Enhancement Is Not in the West," *Conversation,* August 1, 2016.

EPILOGUE: THE BEGINNING

243 *protest movements against CRISPR that have already sprung up:* Alliance Vita, "Stop Bébé GM: Une Campagne Citoyenne D'alerte sur CRISPR-Cas9," www.alliancevita.org/2016/05/stop-bebe-ogm-une-campagne-citoyenne-dalerte-sur-crispr-cas9/; P. Knoepfler, "First Anti-CRISPR Political Campaign Is Born in Europe," *The Niche* (blog), June 2, 2016, www.ipscell.com/2016/06/first-anti-crispr-political-campaign-is-born-in-europe/.

INDEX